About the Author

CAROLINE SELLER MANZO was born in London, brought up in Fiji, and has lived in many different countries around the world. She read classics at Oxford University and now works as a trainer in communication studies. She and her husband, Marcello, have two daughters and divide their time between London, Sicily, and Milan.

Casa Nostra

A HOME IN SICILY

CAROLINE SELLER MANZO

HARPER

NEW YORK · LONDON · TORONTO · SYDNEY

HARPER

Photos on interior page 14 and insert page 14 (bottom) courtesy of Vito Di Giorgi. All other interior and insert photographs courtesy of the personal collection of the author and her family.

A hardcover edition of this book was published in 2007 by HarperCollins Publishers.

HarperCollins books may be purchased for educational, business, or sales promotional use. For information please write: Special Markets Department, HarperCollins Publishers, 10 East 53rd Street, New York, NY 10022.

FIRST HARPER PAPERBACK PUBLISHED 2008.

Designed by Kris Tobiassen

The Library of Congress has catalogued the hardcover edition as follows:

Manzo, Caroline Seller.

Casa nostra: a home in Sicily / Caroline Seller Manzo.—1st ed.
    p. cm.

ISBN 978-0-06-118921-0

1. Sicily (Italy)—Description and travel. 2. Sicily (Italy)—Social life and customs. 3. Manzo, Caroline Seller—Homes and haunts—Italy—Sicily. I. Title.

DG864.3.M36   2007
945'.8093092—dc22
[B]
2006052829

ISBN 978-0-06-137396-1 (pbk.)

08 09 10 11 12 ID/RRD 10 9 8 7 6 5 4 3 2 1

To have seen Italy without having seen Sicily is not to have seen Italy at all, for Sicily is the key to everything.

—J. W. GOETHE

To my family:

MY HUSBAND, MARCELLO, WITHOUT WHOM I WOULD NEVER have come to know Sicily so well, and our daughters, Melissa and Clemency. All three have provided constant support. I also owe special thanks to my brother-in-law Silvio, who supplied the answers to so many of my questions. Finally, my greatest debt is to Maria, my mother-in-law, who constantly encouraged my interest in Sicily and was so proud of her _sicilitudine_.

Contents

ONE   **CASTLE IN THE SUN** • 1

TWO   **CATS IN THE TOWER
AND DOGS IN THE BED** • 15

THREE   **RELICKS AND RUBBISH** • 41

FOUR   **THE CASCINA FIASCO** • 63

FIVE   **GETTING CLOSE** • 73

SIX   **CARPE DIEM** • 89

SEVEN   **MARIA** • 107

EIGHT   **ALL SAINTS AND ALL SOULS** • 131

NINE   **TALES OF THE UNEXPECTED** • 153

TEN   **LA FAMIGLIA** • 175

ELEVEN   **SELINUNTE** • 193

TWELVE   **SICULUS COQUUS
ET SICULA MENSA** • 203

THIRTEEN   **FERRAGOSTO** • 219

FOURTEEN   **CHRISTMAS IN SICILY** • 239

Casa Nostra

MARIA IN THE 1960s

ONE

Castle in the Sun

"CHRIST, IS IT HOT! I DON'T THINK I CAN STAND MUCH MORE of this."

About to melt under the sheer force of the Sicilian sun, my brother Charles takes refuge in the shade of a carob tree. Mopping a freckled brow with the edge of his sleeve, he leans against a conveniently upturned stone drum.

"That's a piece of Greek temple you're sitting on," I point out primly, guidebook in hand.

In order to impress Charles with the wonders of Sicily, I have brought him to one of my favorite haunts, the Cave di Cusa, a tufa quarry where the first Greek colonists of Western Sicily came to hack out massive blocks of calcareous rock for the construction of their temples. Eaten away by the elements, the once gray quarry walls have been bleached to a blinding white by two and a half thousand years of Mediterranean sun. Huge cylindrical blocks are still attached to the bedrock, while finished drums the size of a large dining table, hewn and detached from the stone, lie among the dust and the rubble, ready for transport. The scene bears all the signs of sudden interruption of work in progress, like a building site where the workers, in their various stages of production, have downed tools and walked out, albeit two and a half thousand years ago. Since the why and the wherefore of the abandon have never been explained, an air of mystery pervades this scene of dust and stone and more dust, the dryness relieved only by the odd dwarf palm or wild olive tree sprouting haphazardly from the rocky terrain. Beyond the quarry, row after row of lush green vines merge into a hazy horizon of purple hills outlined against a great expanse of blue.

"How far are the temples from here?"

"About eight miles."

"How the heck did they transport such enormous blocks?"

"Slaves presumably, or mules."

I wave at my brother-in-law to come over and provide an authoritative answer. Silvio, *l'architetto*, is our cicerone, the acknowledged brain of the family. He strides over, immaculate in white linen and Panama hat, only a damp mustache and an unwontedly flushed complexion betraying his discomfort in the heat. Putting me right, he explains,

"Slaves yes, but not mules—oxen. They used oxen to drag the stones all the way to Selinunte." While Silvio delivers a detailed lecture on Greek temple design, Charles, his captive audience, stares glassy-eyed into the crevices of the quarry. Meanwhile, I scan the site for my husband and two daughters. By now I am used to this automatic dispersal on arrival at places of cultural interest. All three are remarkably adept in the art of opting out of sightseeing expeditions. Marcello, identifiable by his khaki safari jacket, is just visible in the distance, half hidden by a clump of bush. Bottom up he seems to be poking at cracks in the wall of the quarry with his walking stick. Lucy, a fox terrier of a certain age, is in the same humiliating position. Obviously convinced that she has a vital role to play in her master's mysterious enterprise, she is barking frantically at his side. What on earth are they doing? They are too far away and the afternoon far too hot to make it worth trying to catch their attention. I look around for the girls. Subjected to relentless sightseeing from a tender age, Melissa and Clemency have learned to deal with such excursions by perfecting a mode of passive endurance; finding a comfortable spot where they can sit the whole thing out is their preferred strategy, while mother and visitors exert themselves. And there they are, stretched out under an olive tree, plugged into their respective music devices. Melissa, the English Rose, fair and freckled like her Uncle Charles, bats crossly at an overattentive fly. Clemency, the Sicilian Jasmine, with glossy dark tresses and an olive complexion as smooth and cool as marble, proffers long-lashed eyelids to the sun, languidly reveling in the sultry afternoon heat. Jack, the Airedale, who has been aimlessly loping about, occasionally lifting a leg on a slab of classical Greek masonry, has finally flopped down to rest beside the "Flowers."

Suddenly Marcello and Lucy emerge from the bushes, covered in dust. Never having seen the point of poring over ancient rocks and stones in the first place, my husband has combined the cultural excursion with an alternative agenda and come up with something far more compelling than ancient history. Triumphantly he brandishes the fruits of his and Lucy's joint efforts—not fruits exactly, but berries.

"Look what I've found: capers! We'll take them home and put them in salt. There's a whole forest of them growing out of the rock. Come and see." Somehow Silvio and I fail to share his excitement. We are all too used to Marcello's habit of treating the landscape as a potential larder. It is a well known fact that capers grow out of rocks and stones, which is why they are often to be found thriving among ruins. Charles, relieved at the break from the lecture on the finer points of Greek architecture, examines the buds and coin-shaped leaves with interest.

"Hmm. Looks exactly like the plant growing in my bedroom at Santa Maria."

"Growing in the bedroom? What are you talking about, Charles?"

"Yep, there's a plant like this with little berries growing right into the bedroom. In fact I can't even close the door to the terrace."

Back at Santa Maria it turns out that capers are indeed growing inside our villa. Charles, who enjoys a comfortable lifestyle as an investment banker in Bangkok, has been in Sicily for just twenty-four hours, but during that time he has endured rigors to which prosperous bankers are rarely subjected. Due to lack of space on the first floor, where we are staying with Maria, my mother-in-law, and Silvio's family, we have had to accommodate him in a less comfortable wing of the villa. Peevishly he points out that apart from the caper invasion, he has noticed, but doesn't choose to examine, unidentified pellets that look suspiciously like rat droppings. Besides, he has had to roam two floors to find running water. Clearly he is not impressed. This is unfortunate, as Marcello and I have been thinking of undertaking a renovation project of his share of the Manzo family home, and we were counting on Charles to provide an outsider's impartial opinion. Despite this inauspicious start, we are still hoping for his approval.

The Flowers retire upstairs to cool off while we find Charles another room. Then, like a photographer aiming to get the subject's best angle, I usher my brother around to the back of the villa in preparation for a dramatic Sicilian sunset. The light from a flaming evening sun filters between tapering spires of cypress trees and Mediterranean pines. The southern facade of the villa, with its patchy, mottled stucco and rows of faded shutters, is washed to a dusky pink in the evening light. A welcome drop in temperature releases a delicious scent of jasmine all around us. The buzzing of cicadas has died out to make way for the call of the resident hoopoe bird, which starts up its nightly chant from an invisible perch high up on one of the towers. At moments like these, Santa Maria seems to be pleading with us to give the project the go-ahead. Surely the scene is captivating enough to seduce even the most hardened financial consultant.

"Well, Charles, isn't it beautiful? What do you think?" From his severe expression I already know the answer before he even opens his mouth. Out come the gobbets of conventional wisdom I had dreaded.

"It's very run-down. Seems to me that a project like this is too big for the two of you. You'd be biting off more than you can chew. If money were no object . . ."

The pathways are submerged in the undergrowth, so to make our way through the garden, we have to battle with a morass of brambles, prickly pears, spiky weeds, and other unfriendly inhabitants of the Sicilian countryside. Puffing and spluttering, his freckles multiplying by the minute, Charles waves his arm in a disparaging gesture at this sorry no-go area purporting to be an orange grove.

"Just look at all this!"

I *am* looking. What he sees as a hopeless, weed-ridden ruin, I see as a typical Sicilian *baglio*—ancient fortified farmhouse, enhanced by a walled garden in the Arab tradition, redolent with the scent of lemon trees. It's true that the walls are crumbling, the pathways are overgrown and the fountain is cracked and parched. It's also true that of the hundred citrus trees planted, only a handful of misshapen specimens have survived years of scorching Sicilian summers without the vital water they need to thrive. Under attack from my younger brother, I feel obliged to rise in their defense,

"Well the garden *has* been abandoned for over twenty years."

"Garden? Is that what you call it? Get real, Caroline. You'd have to start all over again. And let's face it, it not just the garden that's the problem but the villa too. It's practically derelict."

"Oh, come on," I protest, "Don't exaggerate."

"How on earth did it get into this state anyway? If Marcello's family can't afford to keep it up, why don't they just sell it?"

"It's a long story." As I explain to Charles, Santa Maria represents everything that is dear in life to Marcello's mother, Maria. Her father had always promised his youngest child and only daughter that one day the villa that bore her name would be hers. But he died unexpectedly, and his death was followed by the disintegration of his industrial empire and the wealth that accompanied it. In an attempt to revive the family fortunes, in the mid-sixties Maria decided to launch her own wine label. It was an ambitious project involving the conversion of the stables and warehouses of Santa Maria into a wine production center. Unfortunately the project floundered, the banks called in the loans, and Maria had to sell all her assets, including several hundred acres of vineyards.

"But she managed to hang on to the villa itself?"

"Only by selling everything it contained—the furniture, the pictures, the statues. Even the walls were stripped of the coats of arms. She was determined to keep Santa Maria in the family. Long before she became ill, she divided it between her three sons."

"Like King Lear?"

"Yes, and as with King Lear, there was trouble in store."

"Trouble? What kind of trouble?"

"As I said, it's a long story. . . ."

Although Maria had managed to hold the debt collectors at bay, the family fortunes never recovered, and when Goffredo, Marcello's father, died, Santa Maria entered into steady decline. By handing it over to her sons, Maria hoped that they would invest in it and each make a home there. But there were debts still to be paid off. No juggling with figures could defy the grim financial truth—that the outgoings far exceeded the incomings. Without a major injection of capital now, the dilapidation will continue unchecked, and the mounting debts will force the three brothers to sell. If that happens, this could turn out to be our last summer at Santa Maria.

"Yes, but why does the initiative have to come from Marcello? What about the other two brothers? Silvio lives here all the year-round, doesn't he?"

Silvio is indeed the only brother to actually live here. And of the three he is the most deeply attached to the villa. With a summa cum laude in architecture, the equivalent of a starred First from Cambridge, he could have made a career for himself in one of the industrial cities of the north, but he never had any inclination to do so. His *sicilitudine*—passion for Sicily and all things Sicilian—has kept him here on the island, and like his mother, he is determined never, ever to give up Santa Maria. But on a government salary, with a wife and two children to support and his mother to look after, how will he find the funds to halt the steady dilapidation of a sprawling seventeenth-century pile?

As for Pilli—the nickname (meaning "pick-me-up," the insistent refrain of the spoiled toddler) universally used for the eldest brother, Francesco—Pilli the *stravagante*, has inherited not just his mother's charm, but also her extravagant streak. While Silvio in his secure job as a government employee is a man of caution, Pilli has thrown that same caution to the winds to embark on a series of moneymaking schemes, each "investment" more ambitious than the last, and each one leaving him further in debt and more desperate to recoup his losses than before. The problems started when he decided to renovate his share of the villa and make it a home for his ever-expanding family only to abandon the project soon afterwards, when the money ran out. As a result he has become thoroughly disenchanted with Santa Maria, the Family, and Sicily into the bargain, all of which he claims have conspired to drive him into penury. Santa Maria is jinxed. It has ruined his parents and is now going to do the same to the next generation.

The issue came to a head earlier this summer with a telephone call from Silvio. As Marcello put the phone down, his face was contorted with worry.

"Apparently Pilli has found a buyer for Santa Maria. He wants us to go for it. He says it's too good to refuse."

"But Silvio will never agree to that! And what about Maria? What would happen to her?"

"I know, I know, but you can't reason with Pilli." Minutes later Pilli was on the line. Overjoyed at this stroke of *fortuna*, he had already opened a magnum of champagne to celebrate.

"This is a god-sent opportunity. We can't just throw it away, Marcello, *per l'amor di dio*."

With Pilli determined to put Santa Maria up for sale, and Silvio adamantly refusing to budge—not even for billions of lire—the situation was at deadlock. Pilli already had plans about how to invest the capital that would come his way with the sale, and by now he was convinced that his brother Silvio was completely mad. Only a madman would turn down such a lucrative opportunity for sentimental reasons. The more Pilli insisted, the deeper Silvio dug in his heels.

Marcello and I from our base in Milan were not privy to the ins and outs of the battle, but occasionally Marcello would receive an irate call from one of the two brothers, declaring the other *pazzo*, and urging Marcello to intervene. This was an essentially Sicilian drama. If it weren't a true story, it could have been written by Pirandello, Sicily's Nobel-winning playwright, who explored refuge into madness as an escape from reality in many of his plays. Rather than face the *verità*—if indeed there is such a thing—a Pirandellian character will allow that each individual has his own version of the truth, but when someone else's version conflicts with his, he accuses them of madness, which was exactly what was happening now.

Charles reflects on this. I can tell that he agrees with Pilli that the practical solution is to sell. "So what you're saying is that the only way to stop the sale and to maintain your links with Sicily is for Marcello and you to intervene with megabucks. But if Pilli still wants out?"

Impulsive as he is, Pilli is subject to frequent changes of mind, and Marcello is convinced that if we take the plunge and put our money into the villa, he will give up the idea of selling altogether.

"But do you really think that the three brothers will one day live together at Santa Maria in perfect harmony?"

"Hmm. Good question. I hope so. They have fallen out on many occasions in the past, but they always seem to patch up their differences somehow."

"Yes, but three brothers with their wives and families—the whole caboodle—actually living together on the same site. Can you imagine you, me, and our sisters in that situation? It doesn't even bear thinking about. If it is really up to you and Marcello to realize this dream of Maria's, I think you should both think about it long and hard. If you ask me, it sounds like a recipe for disaster."

At supper that evening, Marcello and I make another attempt to convert Charles to our point of view. As we see it, Santa Maria has everything one could wish for in a dream home—above all history and physical beauty. Apart from the investment aspect, what we want most of all is to keep it in the Family.

"Well, if that's the case, why didn't you go ahead and renovate years ago?" We explain that it is only recently that the financial crisis has come to a head. For years we were living on the other side of the world, and the renovation of Santa Maria was a very distant project, something we would tackle in our retirement. Besides, the idea of managing a building project from so far away seemed impractical, to say the least. When we came back to live in Italy, it was not to Sicily, but to Milan, where we had to find somewhere we could actually live. As a result of an abortive attempt to renovate a farmhouse in the countryside outside Milan, a disaster known to family and friends as the Cascina Fiasco, we became painfully aware of the pitfalls of restoring old buildings, especially in Italy. At this point Charles weighs in with a massive dose of common sense.

"Santa Maria will be a mammoth restoration operation, bound to gobble up all your savings and more. Even if you made the basic repairs and dig up the ground floor, where would that leave you? With a piece of a villa hundreds of miles away from where you actually live. How much time would you get to spend here? The major downside is the fact that the property is shared between three brothers. If things went wrong for you financially, would you be able to sell your share independently? And if Pilli insists on selling his, you might be forced to sell yours against your will. Just think of the complications. As an investment, it simply doesn't make sense." When Marcello's back is turned, Charles leans over in a moment of fraternal concern.

"Caroline, it's not just a question of money, is it? After all, this is Sicily, not exactly a prime location. Is this where you want to end up, Caroline? Think about it. Do you really want to live here?"

CHARLES MAY WELL ASK WHY I SHOULD WANT TO LIVE IN SICILY of all places. But he must know that women the world over leave home, family, and friends for foreign climes for the sake of a man. And so it was with me. It was at a Halloween party in London way back in the seventies that my love affair with Sicily began. The room was dark and the air thick with cigarette smoke. Candles entrapped in hollowed-out pumpkins cast an eerie flickering light from zigzag windows. Masked faces loomed close, leered and moved away in a shadow puppet show of cloaks, tails, and witches' hats. Around me couples grooved to the throbbing rhythm of Marvin Gaye's "Grapevine." Peering through my diamante mask, I searched the scene for signs of my escort. What had become of him? Just minutes ago we were in a clinch, part of the rhythmic swaying throng. Ah, there he was—wrapped around a diminutive girl in a Cleopatra wig, his pirate's eye patch hanging jauntily from one ear. Clearly he was not looking for me.

As I felt my way toward the exit, I failed to notice a long trousered leg sprawled across my path. I tripped and would have crashed to the floor had not a pair of grotesque rubber paws grabbed me in the nick of time, pulled me up, and pressed me to their owner's chest. Face-to-face, or rather mask to mask, I found myself staring at a lion improbably sporting a smart green coat. Bits of curly beard protruded from his rubber mask. For a zillionth of a second I was lost for words. Then I let forth, first upbraiding the lion—What the hell do you think you were doing tripping me up like that? Then I repented. Sorry, you probably didn't mean to. Thanks for catching me. The lion did not utter a word. Nor did he let go. Was he dumb or what? A thought struck me. Wasn't that a Loden he was wearing—the Austrian mountain coat that was currently the ultimate fashion item for the young Italian male? If he didn't speak English, I'd try out my elementary Italian on him.

Sei italiano?

If a rubber lion mask could break into smiles, this one would have.

"Non italiano," he corrected me, *"siciliano. Mi chiamo Marcello."* So he could speak—and he had a thrillingly deep voice too. But as I suspected, Marcello didn't speak English, or very little. He was overjoyed to meet someone who could speak his language, however badly, someone he could connect with. I was instinctively drawn to the gentle Lion in a Loden and when he asked me for my telephone number I didn't even hesitate. I scribbled it on the palm of his rubber paw. Only then did it occur to me that we were both still wearing our masks.

Before we parted that night I had a face to attach to the deep voice and tall thin body. Marcello was the perfect match for my preconception of a Sicilian—Hollywood style. The face was long and oval, the eyes light brown and gentle with bushy, arched eyebrows. A slightly receding hairline was generously balanced by a luxurious mustache and a Che Guevara beard. I very much liked what I saw (except for the beard, which eventually would have to go). However, attractive though he was, he was still a stranger who didn't speak my language. And I had given him my telephone number on trust. How could I have imagined then just how huge an impact this impulsive gesture would have on both our lives?

Soon afterward, Marcello took a job in Central America. The pharmaceutical company he worked for happened to have its headquarters in Milan, where I was about to move. For weeks I received cryptic postcards from Panama, Costa Rica, and Guatemala. Sometimes they were of his hotel with an arrow pointing to "my room." *"Auguri, Marcello"* was all he had to say on most of them. On his first trip back to head office he rang me from the airport and asked me rather sweetly to *"andare a ballare."* We never did go dancing. In fact our first date was anything but intimate. Marcello turned up in a borrowed Cinquecento with two giggling women squeezed into the back. There was a third in the front seat. Space was limited, so she ended up sitting on my lap. Inwardly furious, outwardly cool, I tried to work out what his game was. Was the triple chaperone arrangement part of some kind of medieval Sicilian courtship ritual? Or was he putting me to the test as in the *Taming of the Shrew*? From the hilarity of the exchanges between him and the three women, it was obvious that they all knew him far better than I did. At the first opportunity, shamefaced, Marcello whispered into my ear—they were the wives of close friends who had had to attend a company dinner that night—all very last minute. As far as his friends were concerned, Marcello was a *scapolone*, a handsome bachelor at a loose end, so why shouldn't he take the ladies out for dinner? Whether or not he happened to have plans of his own for that evening, they did not bother to find out. And Marcello was too much of a Sicilian gentleman to say no.

I must have passed the Sicilian shrew-taming test with flying colors, for the next day he scooped me up and announced he was taking me to Venice for the weekend.

"Venice is very busy at this time of year. I hope you don't mind, but all I could find was a small pensione. Nothing special, but at least it is clean." Of course I didn't mind, especially when the modest pensione turned out to be Hotel Daniele, *the* most desirable location in the city. One night we were playing sardines in a clapped-out Fiat and the next we were sipping Prosecco in the grandest hotel in Venice. How could I not be captivated by a man so full of surprises?

By the time we returned to Milan, I knew that I had finally met the person with whom I wanted to spend the rest of my life. At this stage it was not at all clear that Marcello had arrived at a similar conclusion. However he did insist again and again that I should meet his family in Sicily and see Santa Maria, the grand, sprawling villa where they lived. He could not take me there himself. He had to go back to Panama, but with the generosity characteristic of the *meridionali*—southern Italians—he put his family and their home at my disposal, assuring me that his parents and younger brother would be only too happy to have a complete stranger as their guest. Of course I wished that Marcello would be coming too, but that was not to be. Nevertheless I had the feeling I was at a crossroads in my life, and so at the first opportunity I took him at his word. I had already met the Man, now I could hardly wait to see the island, the Villa, and above all the Family who, between them, would conspire to take my life in their hands and turn it upside down.

SILVIO AND CAROLINE, 1974

TWO

Cats in the Tower and Dogs in the Bed

IT WAS TWO O'CLOCK IN THE MORNING WHEN THE TRAIN DREW into the station at Mazara del Vallo in the Far West of Sicily. The barely lit platform was deserted, but for a tall, heavyset young man wearing a cap and shooting breeches. He was standing patiently in the dark, as though waiting for someone at this hour of night were a perfectly normal thing to do. I was the only passenger to descend from the train, and as it moved out of the station I was seized by panic. The stranger's height, somber look, and heavy mustache gave him an uncanny resemblance to a member of the British aristocracy recently in the news for disappearing after doing in the family nanny. What was I doing alone in Corleone country at this ungodly hour face-to-face with an absconding nanny-killer?

As soon as the man on the platform stepped forward and introduced himself, my fears were set to rest. Of course this was none other than Silvio, Marcello's younger brother. In any case there was a strong likeness to Marcello. He was taller, stockier, and beardless, but he had the same handlebar mustache, emphatic bushy eyebrows raised in the center like a circumflex, and deep timbre of voice. In my communications with

Marcello's family, somehow I had confused the train arrival times and sent Silvio scurrying off on a hundred-mile trip to Palermo in the early hours of the morning. By the time my train drew in several hours later, he had already given up hope and gone home. To save him a second journey, I decided to make my own way to Mazara. But the train, the so-called *rapido* from Palermo, due in at midnight, failed to live up to its name and was nearly three hours late. I was mortified for the trouble I had caused and for the ungodly hour. But Silvio just beamed at me and dismissed my pathetic attempts at apology with a wave of his hand. He was delighted to meet a friend of his brother's, whatever the hour, *naturalmente*. Gently prising a scruffy travel bag from my grip, he loaded it into a battered Deux Chevaux, opened the passenger door with such a flourish that it nearly came off in his hand, and soon we were lurching and bouncing out of the station.

Given the communication barrier—Silvio didn't speak English and my Italian had not advanced beyond beginners' level—conversation was stilted. *"Scusami, scusami tanto,"* I repeated again and again. When he answered in French, I desperately plundered O-level reserves to come up with *"Je suis desolée."* Still, despite my embarrassment, I was happy, very happy to be in Sicily and not just because of the link with Marcello, on whom I already had discreet designs. I had studied classics at university, and to a classicist, Sicily represented all that was mysterious and exciting. Sicily was the America of the ancient world, where adventurous immigrants arrived from Greece and founded colonies— Syracuse, Selinus, and Akragas to name a few. It was also the battle arena of some of the most exciting episodes of ancient history—such as the Athenian expedition to Sicily, so movingly described by Thucydides, and the terrible defeat at Syracuse, which brought Athens to her feet and lost her the Peloponnesian War. And it was over Sicily that the giant powers of Rome and Carthage came to blows in the First Punic War. Famously fertile and staggeringly beautiful, Sicily was the inspiration for the bucolic poetry of Theocritus and many of Pindar's odes. And didn't the Persephone myth, one of the most haunting that the imaginative Greek mind ever invented, come out of Sicily? Unfortunately my knowledge of the island was a little hazy after the classical period, only picking up in the nineteenth century with Garibaldi and his Thousand starting a revolution, and ending with the movie of *The Godfather* and a dashing Robert de Niro pursuing his vendetta in the very countryside where I found myself now.

From the garbled dialogue that was now taking place between me and my host, I understood that Silvio would be at my side for the whole of my stay. He had taken time off from his job at the *comune* to be my escort. And he had treats in store—Greek temples, baroque churches, Norman ruins, and an Easter picnic. Various friends had been lined up to what? I tried to concentrate but managed to understand less than 50 percent of what he was saying. I had already been twenty-four hours without sleep, and the car was certainly making a lot of noise. I could catch a few key words, but I couldn't actually come out with anything but *"si, si."* He must have been getting the impression he'd be spending a whole week with a nodding idiot.

What with the intellectual effort required to make sense of Silvio's elaborate prose, plus the need to show my appreciation in a language I spoke badly, there was very little energy left over for taking in the surroundings. As we clattered through the outskirts of the town and the buildings thinned out to give way to country roads, the most I could pick out by way of scenery were stone walls or open vineyards. Silvio told me we were only a mile from the sea, but there was no view, and the land was very, very flat. Without lighting of any kind, at this hour it was also very, very black. There seemed to be no landmarks until we completed the last bend in the road, and there it was before us—Santa Maria. The car's headlights revealed the stark outlines of pines and cypress trees, then as we approached the iron gates, parallel lines of palms on either side of the drive. Suddenly, beyond the gates a castle loomed out of the darkness, complete with crenellated walls and Rapunzel towers. It was not a castle in the intimidating military sense, more like a five-year-old's attempt at castle design. Not that I had the vaguest idea at the time of what a Sicilian castle should look like, but it reminded me of crusader castles, or rather a miniature version of one, lifted from the desert and plonked down in a lush Sicilian setting. My pidgin Italian wasn't up to getting any of this across. I blurted out that the *castello* was *bellissimo*. Silvio insisted it was not a castle but a villa, and even so, it was *molto malandata*. It may have been "run-down," but in the darkness all I could see was fairy-tale romance.

Once inside, we climbed a flight of marble steps, at the top of which, to my utter amazement, Silvio's parents were waiting. Goffredo and Maria had stayed up until three in the morning. Not only had they never set eyes on me before, but until only days before, they hadn't even known of my existence. Overjoyed at my arrival, Maria embraced me like a long-lost daughter, and Goffredo bowed and kissed my hand. Too exhausted to take in much of what they were saying or to appreciate the food that Maria had kept warm for me, I mouthed effusive thanks, meanwhile longing for bed. Talking nonstop, Maria ushered me down a long, dimly lit corridor smelling distinctly of drains. I was vaguely aware of a room with a high ceiling, dark cumbersome furniture, and a cold marble floor. A large photo of Marcello puffing on a cigar told me this must have been his bedroom. What happened then has dissolved into oblivion. All I remember now is Maria smiling down on me and wishing me *sogni d'oro*—golden dreams.

I WOKE TO SUNSHINE POURING INTO THE ROOM AND A DAWN chorus doing overtime. By day I saw what Silvio had meant by *malandato*. At best one could say the room donned an air of shabby chic, but there were green mildew stains on the walls and patches where plaster and stone had parted company. A chilly draft came through the broken window shutters. Despite the evident frailty of the antique wooden contraptions, I decided to tackle them anyway in my curiosity to see the garden. The lever was stiff, but eventually it gave way to my exertions with a loud crash, mercifully without resulting in damage. I was looking down on a garden planted closely with trees, but before I could even make out what kind, a great whiff of perfume invaded the room. Even from the upper floor the scent was overwhelming. A forest of glossy green studded with tiny ivory blossoms was enclosed by yellow stone walls, gaping open in parts as though they had been eaten away. Some of the walls were hidden by wisteria floating over them in a great cloud of lilac, others by a creeping blanket of jasmine. The ground was covered with brilliant green undergrowth, dotted with a myriad of yellow flowers. Beyond the far wall, a row of lofty cypress trees shivered in the spring breeze. The scene would have been worthy of a bucolic poet, had it not been for the presence of two moth-eaten-looking mongrels convulsed in enthusiastic barking, apparently overjoyed to receive some human attention.

Having heard the dogs and the battle with the shutters, Maria was now knocking on my door. She had been up for hours and was anxious to show me around. While Silvio and Goffredo slept, she gave me a huge bowl of caffe latte before taking me on a guided tour. Curious as I was to see the rest of the villa, I was even more curious about Maria herself. Marcello had given me a synopsis of her fall from riches to poverty. Even if he hadn't told me, I could see from the state of the house that there was no longer money to spare.

"She is an eternal optimist," he said, "always full of projects and schemes to sell something or other, to write a bestseller, to raise money somehow. She never gives up hope that one of her many projects is about to come off and make her rich again. Be careful," he had warned me. "Don't let her push you around. She is used to having her own way and doesn't take no for an answer.

"*E prepotente, mia madre.*"

If my designs on Marcello were to succeed, Maria would play an important role in my future life, so this prepotente caveat had put me on my guard. Yet, I found her far from pushy. On that initial meeting I warmed to this kind woman, who had stayed up to the small hours to welcome me and was so clearly delighted to have me as her guest. At the time I was totally overwhelmed, but later I realized that the essence of Maria's charm was the intense pleasure she took in the company of others, in my case, of a total stranger. With a clear pink-and-white complexion devoid of lines or wrinkles, green eyes, and a rosebud mouth, she had the aura of a prewar movie star. Various portraits placed around the sitting room confirmed this impression. One photograph in particular of the young Maria posing confidently for the camera, wearing an

ermine stole and diamond necklace, evoked the glamour of a bygone age. That day, dressed theatrically in a bright flowery skirt, flashy costume jewelry, and a gypsy shawl, she could have stepped off the stage of a production of *Carmen*. Not exactly what one expected to see on the lady of the manor, not even one who had come down in the world. The Sicilian—or was it only this family's—penchant for dressing up had not escaped my notice. On the previous night, Silvio had donned the tweeds of an Edwardian squire on a Sunday shoot, and Marcello, when I first met him, had been dressed as a lion.

I knew from Marcello that his mother was an extrovert, but never in my life had I met anyone as outgoing as this. She was full of questions, not about why I was traveling alone, for that was obvious—I'd come all the way down the leg of Italy to meet her, of course—and full of compliments.

"*Come sei bella*—not as slim as I was at your age, but all the same *molto bella*."

I took the reference to my solid build in good part, for having lived in Italy for several months, by then I had learned that personal remarks of this kind were meant as expressions of interest rather than criticism. She talked to me as though we were sisters who had been separated since birth and she had a life history to cover. I answered the barrage of questions as best I could, preferring to listen to her rather than recount a life story of my own. The fascinating rags-to-riches story she told was already familiar to me in Marcello's version—how her father, Luigi Vaccara, had started life as son of a sharecropper and ended up as "the Rockerfeller of Sicily." Maria was as proud of her father's humble origins as she was of his achievements. In his early twenties he had built up his own wine industry. From wine he branched into fishing and shipping, launching the first fleet of engine-powered fishing trawlers from Mazara. This was in the 1930s, the age of Fascism and Mussolini's Black Shirts. I dared not ask her how far Luigi had had to toe the Fascist line, but on a later occasion she showed me a photograph of the family in those early days: Luigi's five children all dressed in black, the four sons in descending order, with a sullen little Maria at the end of the line. This is the only photograph I have ever seen of Maria looking unhappy.

"Papa was Podestà, Mayor of Mazara, and a close friend of the Minister of Aviation, who even made available military planes to fly him back and forth to the Vaccara wine factory in Tripoli." She must have seen the alarm on my face at the mention of a Fascist connection.

"No one held it against him after the war. Everyone was a Fascist in those days," she explained. "You had to go along with it to survive. There was no choice."

Before we left the room Maria showed me a photograph of herself holding Marcello and Pilli in her arms. Aged two and four, respectively, they were dressed in girlie outfits, their dark locks curled into

ringlets. Somewhat aghast at the effects this kind of parenting might have had on the Sicilian male, one of whom I was planning to marry, I asked, "Was it normal to dress boys like that in those days in Sicily?"

"Well, they looked so sweet and I always wanted girls." I was beginning to see what Marcello meant by prepotente.

"Did you know I went to boarding school when I was only six years old?" I was surprised. I had been told that boarding school had never been popular in Italy; it was a last resort for dealing with misfits or served as a crammer for the rich-and-thick. Most Italians with money are happy for their children to rub shoulders with the less privileged. Otherwise they send them to private day schools run by the Church. Boarding for its own sake is regarded as cruel and unusual punishment for children and parents alike, especially parents, who are generally loath for their little birds to fly the nest. This reluctance to let go still applies even when the fledglings mature into middle age. So what was Luigi doing by sending his baby girl away? I was about to commiserate when she cut me short, full of enthusiasm.

"*Si, si, La Santissima Annunziata di Poggio Imperiale a Firenze.* Haven't you heard of it?"

I had to confess I hadn't.

"You must have heard of it. It's the most famous girls' school in Italy. Even the Queen went there."

"The Queen?"

"Not your Queen, The last Queen of Italy. Look!"

She pulled out a photograph dated 1922 and signed "Maria José of Belgium." I was eager to see more photographs, but Maria had other things on the agenda.

"I'll show you the photo albums later. Come into the kitchen and meet Antonietta. She's making *pasta al forno* in your honor."

The kitchen wore the same look as the rest of the house, but with more of the shabby and less of the chic. The walls were brown with smoke and the ceiling completely black, presumably the result of accidental fires. In the chaos on the central table there was a pile of the reddest tomatoes I had ever seen. Antonietta was tiny. Hunched over the table with brown wizened face and hands, hair drawn tight under a headscarf and apron tied over a shapeless black dress, she fitted the stereotype of Sicilian widow to a tee. She grinned at me mischievously, exposing gums almost bereft of teeth. How old could she be? Far too old to be working, surely. Maria read my thoughts.

"She's only about my age (which I took to be sixtyish), but she doesn't know exactly when she was born. They didn't keep official records of births in those days. She's had eighteen children. That's why she looks old. Actually she's as strong as an ox."

Out of Antonietta's earshot, Maria told me she had been working for the family for many years. At the time of their worst financial crisis, she had gone without wages. Maria was so grateful that she had given her a plot of land so that Antonietta could build her own house. Could this tiny little woman really have had eighteen children? Catching the word *figli*, Antonietta began gabbling away in dialect, which Maria had to translate. Seven of them had died, the rest had left home except for two sons, grown up now but still unmarried. She had made an early start that day, as she had to leave by eleven to get lunch for these two grown sons. Then she would be coming back to the villa to wash up after lunch.

Leaving Antonietta to get on with the baked pasta, Maria was keen to show me the rest of the house. She explained that the family had lived upstairs ever since her mother had become an invalid and could no longer manage the stairs. To be able to accommodate Nonna Francesca, the main kitchen had been moved to the first floor, to what had once been a breakfast room. The upstairs sitting room became the main reception room, and the rooms downstairs were abandoned. So instead of relegating Granny to her sickroom, the whole family had moved upstairs in order to keep her company. By the time her mother died, Maria could not afford to reconvert the villa, so the family remained in what was for all intents and purposes, an upstairs flat. Maria was optimistic that they would one day use the ground floor again.

"But of course we'll move the kitchen and dining room down-stairs again, as soon as things look up." (As soon as things look up? Why would they, and on what basis did she make this assumption?

But these were not questions I could put to Maria at the time.) She wanted to show me the courtyard by daylight, so we descended the marble stairs to the ground floor. In the spring sun, the towering walls presented a variegated palette of hues from butter yellow to dusky rose, that is where they were visible beneath the foliage. Ivy, Virginia creeper, and vigorous bougainvillea had staked their claim as far as the upper terraces. Only the twin towers were free of overgrowth, the plaster glowing pink in the sunlight like a ripened peach. Arched doorways to the ground floor rooms flanked the double staircase on both sides and directly opposite, below the ramparts were a series of porticoes. To the right, the third side of the square was formed by the massive walls of a warehouse. To the left a vaulted porch, led to a grand wooden door, the front entrance. Above the porch was a square turret with iron bell and a balustrade lined with pink geraniums. I gasped out my appreciation, *"E veramente bellissima!"*

Maria sighed. "If only you could have seen it in my father's day!"

Shutters of rotting wood, broken panes, and plaster stained with damp bore witness to the fact that the courtyard had seen better days. An octagonal well in the center gleamed white in the spring sunshine. Maria proudly showed me her initials—MV, for Maria Vaccara—carved in relief on the marble. Otherwise the only ornaments were a collection of broken amphorae sprouting cacti and a terracotta one-armed Bacchus presiding over the courtyard from a niche in the wall. A dog of a breed I didn't recognize lay comatose in the shade of an arched portico.

"That's Greta, Silvio's hunting dog. She's a *bracco italiano. Un cane molto aristocratico.* She's also our guard dog." The aristocratic hound slowly lifted her head and took a bored look in our direction. Then, saggy-jowled and droopy-eyed, she lowered it and sank back into torpor. She may have been saving her energy for more interesting visitors, but I felt sure her lower class colleagues shut up in the back garden would have been more effective on guard duty.

And what happened on the ground floor? Were all the rooms abandoned when the family moved upstairs? Apparently one end had been converted into a small apartment for Pilli's family. This was the oldest part of the house, once a granary where the original vaulted ceilings and high windows had been retained. I knew that Pilli, the only married son, was a civil engineer who spent much of his time in Africa. Carved wooden masks, straw hats, and African prints on the white plaster walls added a bizarre tropical dimension to the former monastery. A family photo showed a chubby little black girl with perfect teeth.

"That's Lul, my granddaughter, Pilli's eldest. He found her in an orphanage in Somalia and brought her back." On impulse and without even consulting his wife, it turned out. A bit reckless of him, even his doting mother had the grace to admit. "*Così è Pilli*—that's Pilli." She sighed.

The rest of the ground floor was hardly used, except by Silvio, who had turned the large room opening onto the back garden into an architect's studio. From here we ambled through an open-spaced living room divided only by a succession of archways. The ceilings were vaulted, and shuttered windows along the wall opened onto the garden of citrus trees I had glimpsed from the upstairs bedroom window. A quick glance at the profusion of papers, paints, canvasses, magazines, rolled-up sketches, and piles and piles of books suggested that tidiness was not a strong point in the family. Maria showed me the chapel, bearing the invitation *Venite adoremus* over the door. This was where Mass used to be held for the farmworkers during the wine harvest. She was hoping her sons would get married here at Santa Maria one day. What about Pilli? Did he get married in the chapel? No, Pilli had disappointed his mother by refusing to have a church wedding altogether. But still, she supposed, she might one day persuade him to do so. It was never too late.

From the chapel we crossed the courtyard to the warehouses. Nothing of interest there, according to Maria. Perhaps they reminded her of her doomed foray in her father's footsteps. Marcello had already told me about the disastrous launch of the Maria Vaccara wine label. All that was left of the misconceived business enterprise were rows of huge concrete vats erected in place of the original barriques. These were hideous, now quite useless, and almost indestructible. They would have to be blasted out with explosives, according to Silvio and Goffredo, who had now joined us.

Goffredo, like Silvio, was dressed in tweeds, a Sicilian version of the country gentleman. Doffing his flat cap, he addressed me with the elaborate courtesy of a Jane Austen hero, enquiring politely after my night's sleep and my physical well-being. He was concerned that my clothing—jeans and jersey—might be inadequate for Sicilian spring weather. Both father and son sported the lavish Manzo mustache, Goffredo's pepper and salt, Silvio's thick and black, but otherwise there was no physical similarity between Goffredo's compact dapper figure and his son's tall, imposing frame. In Maria's dominating presence, both appeared to be men of few words, that is until the subject of wine came up. Maria claimed to be *astemia*—one who doesn't drink at all—but she was still prepared to vouch for the excellence of the wine of Santa Maria. *"Buono, buona,"* she declared it, *"Eccezionale."* Empty bottles lay in piles on the floor of the abandoned warehouse. Maria picked one up and scraped off the dust to show me the label. There was a sketch of the villa, with its twin towers and the turreted blocks of the warehouses behind the main building. *CASTELLO DI SANTA MARIA Amarasco Riserva—Proprietaria di Maria Vaccara, Mazara del Vallo—*was printed in Italian, and then an English word *Italy*, just to give the wine an international touch. Goffredo explained that in the heyday of the Maria Vaccara wine business, they had had international clients. The best customers were the Irish. He was pleased to hear I had Irish blood on both sides of my family and told me he knew Ireland well. He had often been there to promote Maria's wines in the sixties. He proceeded to recount how he had lost a wallet in Dublin containing three hundred pounds in cash, an enormous sum in those days. When he got back to Mazara there was a parcel waiting for him. Inside was his wallet stuffed with the money. The sender had written, "I found this in the street and thought you would want to have it back. Yours sincerely" with name and address provided.

"That could never happen in Sicily. We don't have your tradition of honesty." Goffredo had definite ideas about our national characteristics, English and Irish being conveniently interchangeable in his mind. His assumptions predated the advent of the multicultural society and seemed to be derived mainly from literature. Just as Italians continue to refer to anthracite gray—almost black—as *fumo di Londra*, as if London were still filthy and foggy as in Dickens's day, Goffredo's ideas were touchingly outdated. He endowed the sons and daughters of Albion with an assortment of admirable qualities that ranged from the endearing to the heroic. If he had ever spoken English, he had by then forgotten it, but he had read Italian translations of *Stalky and Co., Three Men in a Boat,* and much of P. G. Wodehouse and Agatha Christie—all of which clearly demonstrated that my country of origin was populated by humorous, punctual, tea-drinking, umbrella-carrying, cricket-mad, dog-loving, green-fingered, fair-playing, scrupulously honest *gentlemen.* Admittedly this was before the appearance on the world scene of the 'ooligan, who has since seriously tarnished the image of *uomo britannico.* As I was determined to be a *persona grata* to the Manzo family, Goffredo's magnanimous version went unchallenged, and I blithely acquiesced to membership of a mythical race.

Goffredo gently brushed off the dust that fell on my shoulders in the warehouse and took my elbow to steer me clear of debris as I was about to venture outside. By now I was captivated by the beauty of the run-down castle and overwhelmed by the kindness of my hosts. Right from the beginning I had detected something that smacked of courtship in their attentions toward me. That week the weather was positively balmy, the temperatures never dropping below 70 degrees

Fahrenheit, but even so, Maria would warn me of the dangers of *colpi d'aria*—chills—and procure a shawl, which Silvio would solicitously wrap around my neck and shoulders, while Goffredo looked on with evident approval. All three plied me with food and drink whenever the occasion arose, anxious that I should sample the best that Sicilian cuisine had to offer. At first I attributed all this to the famous Sicilian hospitality. Yet there was a niggling doubt at the back of my mind. Just how had Marcello presented me to the family? Had he made it clear I was his girlfriend, or was I supposed to be just a friend? Did the warmth of the welcome reflect my prospects as Marcello's *fidanzata*, or was this how a Sicilian family usually treated stray foreign girls who turned up on their doorstep?

These were crucial questions, but they were to remain unanswered in those days of limited intercontinental communications. I could hardly use the telephone to call Marcello in Panama, and then, with Maria standing by, ask him if I was his official girlfriend or not. And what about Silvio, who had taken time off work in order to act as my host? Surely that was overstepping the obligations of hospitality. And Silvio could not have been a more perfect host. Generously, he took me under his intellectual wing, showing me the architectural treasures of Mazara, from the Norman to the baroque—he was engaged in restoring the facade of a Jesuit church at the time—going to great trouble to put each church or palazzo into a historical context. When he discovered my interest in Magna Grecia, he duly took me to the Greek temples at Selinunte and Agrigento. Sometimes he brought a friend along, one of whom was Vito, a budding photographer and Anglophile, who nursed ambitions to become Sicily's answer to David Bailey. Vito took dramatic pictures of me posing by a temple on the acropolis of Selinunte, gazing out to sea, hair lifted by the wind.

"Like a statue of Minerva," Silvio gallantly suggested.

Thinking of my waistline, steadily expanding to accommodate the generous Sicilian meals, "More like Juno," I replied.

On the last day of my visit, Maria took me up to the terrace that looked out from the front of the villa onto the avenue of palms and bougainvillea. Beyond the gardens of the villa, vineyards stretched as far as the eye could see.

"*Questa terra era mia, tutta mia, ma ora è tutta venduta.* Once this was all mine, but it's all been sold." How many times would I hear that phrase in years to come? To her great chagrin, much of it now belonged to families who had once worked on the Vaccara estate. But she hadn't brought me out to admire the view or reminisce on better days. There was something else she wanted to show me. At the far end of the terrace she opened the door into the top of one of the Rapunzel towers. There was a squeaking and a scratching and suddenly a torrent of feline life poured out onto the terrace. A host of skinny little pussycats dived at the bag of scraps she had brought to feed them. The smell was none too pleasant. As she changed their water she sighed, "*Povere bestioline*—poor little beasts. Somebody has to look after them." Then enlisting my solidarity, "These are my *piccirriddi*—my little creatures—and this is my secret. Promise not to tell anyone! If Goffredo finds out, he'll have them all drowned."

It was no use trying to convince Maria that sometimes you have to be cruel to be kind. Evidently Goffredo was exasperated by her love of bestioline. The two elderly mongrels in the orange grove were the last of a long line of adopted strays that she used to take in until Goffredo finally put his foot down. She had evaded his veto on any extension to hospitality toward bestioline by getting Antonietta to cook pasta for the community of stray dogs who would be waiting at the front gate every morning.

But Goffredo found out about that too. Now Antonietta cooked the dog food out of sight in the makeshift kitchen on the ground floor—the dog kitchen, as she called it—and Maria took it by car to the municipal rubbish bins where the gang now hung out. How long would it be until Goffredo came up here and discovered this new lot of bestioline? Most husbands would draw the line at a tower full of cats, multiplying by the day. I simply didn't have the courage to say I agreed with Goffredo. I kept Maria's secret, but when I returned to Santa Maria a year later, the cats had gone.

When it was time to depart, Maria, genuinely desperate for captive female company, literally begged me to stay, then, once she had come to terms with the fact that I was actually leaving, promised to visit me in Milan. Silvio, now comfortably established in the role of platonic friend and mentor, had already accepted my invitation to stay with my family in England, while Goffredo, as loath as his wife to see me go, clasped my hands in his as he said good-bye and wished me *tante cose*—many (good) things, the courteous valediction of the Sicilian gentleman. At the very last minute Maria thrust a fat envelope into my hands. It contained the manuscript of *The Dons*, her latest novel, which she wanted me to take to one of the major publishing firms in Milan. It had been published in English, thanks to the intervention of an English writer friend, but she had never managed to convince a major Italian publisher to take her on. Having sold very few copies in Britain, *The Dons* had not launched her on the literary career of her dreams, but her success in getting into print in English had fanned an ever-present flame of hope. Her failure to make the bestseller lists in Italy she attributed to a lack of *raccomandazione*—the right strings to pull. But now fortune had presented her with an *inglese* with many business contacts in Milan. My job as a humble language trainer had been exalted into something far more important in her mind's eye. Whatever my misgivings, I was powerless to refuse, so I accepted the bundle, or rather the burden, and promised to look for a publisher. It was my debut into Sicilian mores. I had been on the receiving end of unlimited hospitality, generosity, and affection and it went without saying that sooner or later I would be expected to show my appreciation by doing my hosts a favor in return.

Eventually it dawned on me that Marcello had organized my mission to Sicily in the expectation that Santa Maria and the Family would do his courting for him. Bowled over by the hospitality of Maria and

Goffredo, dazzled by Silvio's learning and intoxicated with the sheer beauty of the place, I would fall—*plop!*—like a ripe fig, into his, Marcello's waiting arms. Which is more or less what happened. If that was indeed his strategy, then I have to admit it worked. By the time I was on the night train back to Milan that spring, I was head over heels in love with the whole Family, the Villa, and all things Sicilian.

The feedback from Santa Maria must have been positive, as from that point on the tone of Marcello's letters became more intimate. Gone were the cryptic postcards. Instead he took the trouble to write proper letters. As I opened them I was frustrated by his oddly sloping handwriting, which was so very hard to read. (Only later did I learn just how much effort these letters had cost him. Left-handed by nature, he had been forced to write with his right hand as a child by his governess. The Signora Angelo had grimly supervised his early attempts to write with his left hand tied behind his back, thus ensuring that writing would be a torture for him for the rest of his life.) Before long the missives had developed into full-blown love letters. I responded in my best Italian, my efforts expressed and polished up with the expertise of an Italian girlfriend.

Although Marcello and I managed to get together for a glorious summer holiday in Mexico and a week in London to celebrate our Halloween anniversary, the frustration of living so far apart was mounting, and it was becoming all too clear that a love affair fragmented by long separations and conducted at a distance of six thousand miles would be hard to keep up. Besides there was the jealousy component. Marcello was worried that I might meet someone else, and I had discovered that I was not the only female interested in the handsome Sicilian bachelor. One ex-girlfriend in particular was pursuing him relentlessly, brazenly pretending to be his secretary if I happened to answer the phone.

"We can't carry on like this!" Marcello exploded in one of our rare long-distance telephone calls. "Give up your job in London and come and live with me." Yes, well, I could have done that, but it was not clear what my role would be if I quit all and moved to Panama. How long would he be staying there, and how would I spend my days while he was at work—hanging out at the club, playing bridge and golf, or sunbathing at the pool? What about the Anglo-Saxon work ethic? How would he introduce me to people—as his mistress? What exactly did he have in mind?

In the end the decision was taken out of my hands. Marcello's time in Panama was almost up, and he had applied to an Italian company for a job in Milan. Meanwhile we arranged to spend Christmas together in Santo Domingo. I arrived laden with presents for the man I loved, only to find that Marcello had nothing for me.

"Sorry, I hadn't realized. We don't give each other presents for Christmas in Sicily," he apologized. "Why don't we go downstairs and get you a present now?" What? Go down to the hotel lobby on Christmas Eve to buy a present for myself? If Sicilians didn't do presents, what on earth did they do at Christmas? I was dangerously close to a sulk. Sensing something was wrong, Marcello groped in his pocket and pulled out a small brown package.

"Hey, this is for you."

When I had unwrapped the last layer of paper I was holding a large, clear-cut emerald in my hand.

"Marcello! Are you crazy?" I reeled at the idea of how much he must have spent on a one-karat emerald. "Where did you get this?"

"Oh I picked it up in Colombia," was the reply. "I was going to have it set in a ring—your engagement ring—but you might as well have it now." My engagement ring? Wasn't this jumping the gun somehow? Neither the word *engagement* nor the word *marriage* had passed his lips so far.

"Well, you know that job I applied for? I got it, but it's not in Milan anymore. It's in São Paulo, and if you want to come to Brazil with me, we have to get married first. Otherwise you won't get a residence permit. *Hai capito?*

I got it. This was not the romantic proposal I had been dreaming of, but it was typical of Marcello, always full of surprises. We had spent merely seven weeks in each other's company since we first met, and now he was all set to take two huge leaps in the dark—to get married when we still hardly knew each other and emigrate to a country neither of us knew anything about.

While I reveled at the reckless turn our lives had suddenly taken, Marcello was in a hurry to get the formalities over and done with. The idea of a white wedding with all the ceremony and fanfare did not appeal to either of us. Besides, as Marcello pointed out, there simply wasn't time. His new employers in São Paulo were expecting him within weeks. We had to get married and *subito!* A register office ceremony at my parents' home in Beaconsfield, a quiet commuter town outside London, seemed the most practical solution. The Sicilian family would be sorely disappointed, especially Maria, who had no doubt been laying wedding plans for weeks, if not months. However, before the wedding in England we would go to Sicily together to celebrate the news with the family. Although Maria would not realize her dream of a wedding in the chapel at Santa Maria, she would throw a prenuptial party for us and, most important, she would have a whole week in which to introduce me to the Vaccara family as her official daughter-in-law.

Whereas I was fortunate in that I had mastered Italian by then and the Sicilian family had already taken me under their wing, Marcello still had dragons to slay. He would now have to present his credentials to a bunch of eccentric, opinionated, highbrow, joke-obsessed adults—the Seller family. (Charles was in Bangkok at the time, so Marcello was spared the potentially most intimidating encounter of all.) And he would have to face this challenge with a serious handicap—lack of English, as his skills in that department had progressed very little since our first meeting. I expected him to cower in fear at the formidable lineup. He did no such thing. Very much aware of the importance of Family, he was determined to make a *bella figura*—a good impression. He came to the first meeting dressed to kill, his lanky form set off to its best in a tweed suit from one of Milan's smartest tailors and sporting a suitably discreet tie and traditional English brogues. The nasty guerrilla-type beard had long since gone on my insistence, and he now proffered an extremely handsome, clean-shaven profile to his prospective in-laws. Armed with Cuban cigars for my father, a string of pearls for my mother, and Colombian rings and bracelets for my two sisters, he charmed them all. Handicapped by the language barrier, but longing to endear himself to his new family, he adopted an alternative strategy—the Italian clown act. Family outings were his big moments. He would insist on taking the wheel, seizing the opportunity to treat a captive audience to a show of wizardry on the road, masterly accomplishing last-minute overtaking of unsuspecting drivers, subtly speeding in the inside lane, executing a stunningly illegal U-turn in front of Buckingham Palace, and performing numerous miraculous parking feats. Much of this was accompanied by comic body language—fist shaking, enthusiastic waving, and whoops of joy—all of which had the captive audience gasping in horror and admiration by turns.

Meanwhile, to compensate for his lack of conversational skills, as he drove along, Marcello would read out road signs, advertisements, and poster slogans in a thick Al Capone accent, which had my sisters, Victoria and Emma, permanently doubled up with laughter in the back of the car.

The Sellers were simply loving the new addition to the family. Meanwhile Marcello himself was having a whale of a time. He was full of enthusiasm for everything English.

"*Che bel verde!*" he would exclaim time and time again.

"Only achieved by months on end of rain," I pointed out.

Having seen London already, he had been duly impressed by the glamorous tourist attractions of parks, museums, theaters, trendy shops, and the sheer grandeur of the capital. Now, based in suburbia, he was getting a taste of what it was actually like to live in England as a native. Despite the Italian prejudice against English food, he came out in favor of my mother's steak and kidney pie, asking for second, third, and on one occasion, a fourth helping.

Marcello was quick to notice that the way to my mother's heart was through her dogs. Immediately he made friends with Toby and Bounce, the family Jack Russells, stroking and patting them whenever my mother was present. These two elderly, spoiled terriers made a habit of getting into bed with my parents in the morning, burrowing down under the duvet to the foot of the bed, as though going to earth after a badger and simply lying there at their pleasure. They would then move from bedroom to bedroom in order to extend this unhygienic welcome gesture to any one else who happened to be staying in the house. As I was quite used to this kind of behavior, I had forgotten to warn Marcello, so when the invasive pair arrived in our bed one morning, I just let them get on with their usual routine. Marcello, rudely woken by the wetness and warmth of pulsating animal life sharing his intimate space, leaped out of bed with a scream.

"Get them out of here!"

"Marcello, I can't do that. This is their house. They're just being friendly. And besides, if you shoo them out, my mother will never speak to you again."

Marcello got back into bed, carefully drawing his knees up to his waist in order to avoid unwelcome intimacy with canine life. That night I overheard him on the phone to Maria.

"Every thing's fine. *Absolutely fine* but, you know something, Mamma, the English are very strange. You'll never believe this but in Caroline's family *they sleep with the dogs!*"

Having overcome his aversion to dogs between the sheets, Marcello now had to face the ordeal of the wedding ceremony. Not that he was worried about the act of betrothal itself, at least I assured myself he wasn't, more like he was simply terrified at the idea of getting married *in English*. In an attempt to allay his fears, I organized dedicated training sessions. Every night for a week we went through the words he would have to repeat.

"I, Marcello Edoardo Manzo, do take Caroline Anne Seller as my lawful wedded wife . . ." By the night before the great day he seemed to have learned his lines by heart. The wedding was to be a small affair with just my close family and a few friends in attendance. The ceremony at the local register office would be followed by a simple lunch back at my parents' house. But somehow on the day itself, things began to go wrong. Just before the ceremony was due to start, the registrar approached Marcello, clapped him on the back, took aim, and fired a barrage of incomprehensible language. "Lesgettamoney-outtaway" was what it sounded like to Marcello. Sign language would not help him out of this one, but luckily my father came to the rescue and handed over the required payment. The ceremony was delayed for half an hour as we waited for Emma and her boyfriend to turn up. With the late start Marcello began to get nervous. When it was his turn to speak, he forgot his training lessons and in a faltering voice announced to all assembled:

"I, Marcello Edoardo Manzo, do take Caroline Anne Seller as my *awful* wedded wife."

Giggles from the front row. My sisters were in paroxysms at one more comic performance from the buffoon from the land of the Godfather. I turned around to give them an elder-sisterly look, meaning please shut up. At that point the wedding ceremony launched into farce. When it was time for Marcello to put the wedding ring on my finger, it seemed to be too small. He fumbled, made two nervous attempts, then gave up. Encouraged by the discreet sniggering coming from the Seller ranks, Marcello, once again in buffoon mode, handed the ring over to me.

"I can't do it. Put it on yourself." In desperation I gripped his hand to make it look as though it was he and not I who was slipping on the ring. The ring was on and finally we were pronounced husband and wife. *Phew!* The ordeal was over. But no it wasn't. Not quite yet. The registrar was asking Marcello something, and again he didn't understand. When it was repeated for the third time, I had to translate sotto voce.

"He's telling you to kiss me."

"What now? In front of everybody?"

"Yes. Now. Go on. You have to!"

With that Marcello closed his eyes in resignation, gripped my shoulders and gave me a demure peck on the cheek.

PILLI AND MARCELLO AS TODDLERS WITH SILVIO

THREE

Relicks and Rubbish

CHARLES IS NOT THE ONLY ONE TO NEEDLE US ABOUT THE location of our dream home. As my brother has so clearly pointed out, western Sicily is not exactly prime location for real estate investment. Fellow Italians smirk and chuckle when they find out where Marcello comes from,

"Mazara del Vallo? Isn't it in Africa?"

Santa Maria lies just two miles north of Mazara, a city of fifty thousand souls on the very western tip of Sicily, less than a hundred miles from Cape Bon in Tunisia. But apparently North Africa, as opposed to fashionable Africa, Karen Blixen's or Nelson Mandela's Africa, is uncool. The hackneyed quip associating Sicily with her southern neighbor goes back at least as far as the Romans. Trimalchio, the nouveau riche magnate, whose lavish feasts were immortalized by Petronius, would joke that he was thinking of adding Sicily to his estate in Apulia to save him the trouble of leaving home when he felt like visiting Africa.

While the Africa joke is repeated ad nauseam and no more intended as a compliment nowadays than it was two thousand years ago, western Sicily—Palermo and the province of Trapani—including

Mazara del Vallo, feeds the national prejudice by regularly providing the media with stories of drug hauls and Mafia arrests.

"How come all the bad press comes from places around here?" I once asked Marcello. "Don't they ever get up to anything on the other side of the island?"

"There's a difference between them and us," he replied, "Here in the west we are '*sperti*.' "

"What do you mean 'sperti'?"

Marcello's answer was to pull down the lower eyelid of his right eye with his forefinger. I got the message. "And the east?" "Sicilia '*babba*,' " said Marcello. "In the east they're all fools."

He was referring to a saying well known in western Sicily (and I have never heard it repeated by anyone from the east). *Sperta*—smart— also implies dangerous and *babba*—stupid—means naive and unthreatening. Although for centuries the island was divided in three parts for administrative purposes, the real divide is a cultural one between east and west. The east, where the Greeks first arrived, and the Byzantines resisted the Arabs the longest is the Sicily of Taormina and Syracuse, the relatively Mafia-free, progressive Sicily, the "good" Sicily that attracts the tourists. The west, on the other hand, is disreputable Sicily, where the Arabs arrived from Africa and where the Norman conquerors met the fiercest Muslim resistance. In the last thousand years the west has developed a tradition of rebellion and contempt for authority that has made it the natural home of the Mafia.

Mazara, in the heart of Sicilia sperta, with its proximity to Africa and its Mafia associations, has never been an obvious magnet for tourists. On the other hand, a distinguished British writer sees in Mazara not Africa, but the south of France. In a recent book on Sicily, Norman Lewis describes Mazara as "one of the most charming Mediterranean towns. On all counts Mazara is to be admired for its elegance and calm. It is a ghost of a corner of the French Riviera of old, with a promenade and palms." The British were not always so fulsome in praise of Mazara del Vallo. In the early nineteenth century, an Englishman, an Irishman, and a Scotsman were all three equally dismissive. Augustus Hare, an eminent Victorian, called Mazara "a miserable episcopal town." A certain Irish Lord Ossory recalled a proverb whereby every house in Sicily contained a priest and a pig. After a trying ride on mule-back through Mazara, the "old saracenic town, with embattled walls and towers," he noted in his diary, "We encountered thirty-four ecclesiastics on our ride through it, but I took no heed of the pigs." According to John Galt, the nineteenth-century Scottish novelist, "The town, like all Sicilian towns, is, for its extent, abundantly showy at a distance. The fortifications, being in the oldest and most obsolete style, have a formidable aspect, but nothing more. The town within is a collection of relicks and rubbish."

Two hundred years later the rubbish is still there, mainly concentrated in the outskirts of the town, but I have to disagree with Galt, as the relicks are truly magnificent; in fact Mazara boasts one of the most beautiful baroque piazzas in Sicily. Charles does not believe in sightseeing, protesting, "Calm down, Caroline. You're so driven. I'm perfectly happy to relax and drink in the atmosphere." But I am determined that my brother is going to appreciate this architectural oasis in Marcello's home town. In the many years that we have been coming to Mazara together, we have shown the town to visitors so many times that Marcello, never keen on culture or physical exertion, especially not when the two are combined, has perfected the whistle-stop tour conducted from the comfort of an air-conditioned car. Today he proposes a tour on wheels—a leisurely drive down the palm-lined promenade, a quick spin around the main piazza, a zoom-in to a couple of baroque masterpieces, followed by a glimpse of the fishing trawlers moored in the port at the mouth of the river, and ending up with an *aperitivo* beside a cousin's pool. Charles, equally reluctant to expend valuable energy on useless endeavor, rejects my proposal to explore the town on foot and heartily concurs with Marcello's proposal.

How can I leave my brother in ignorance of the town's fascinating history? Mazara (the name is thought to derive from a Punic word meaning "rock") was founded by Phoenicians as a trading post in the eighth century BC. Soon after it was founded, the Greeks of Selinunte took it over and made it their emporium. Lying on the boundary between the territory of Selinunte and her enemy, Segesta, the port became a bone of contention between the two cities, and in 409, the Carthaginian general Hannibal, ancestor of his more famous namesake, razed it on his way to attack Selinunte. Inhabited by Greeks, Romans, and Byzantines, Mazara's apogee came with the advent of the Arabs in the ninth century. The emirs divided Sicily into three "wali's" and the western and largest division, the Val di Mazara, took its name from this town. The Arab occupation made a strong impact, imposing a layout on the heart of the city which has lasted to this day—that of a medieval kasbah, a maze of twisting narrow lanes, culs-de-sac, and tiny courtyards that Sicilians call *occhi nel cielo*—peepholes in the sky.

For an idea of what the twelfth-century town looked like we have the testimony of Abu Abdullah Mohammed al-Edrisi, an Arab scholar at the court of the Norman King, Roger II. He wrote a geographical treatise of the known world with the splendid title, "The Pleasure Excursion of One Who Is Eager to Travel the Regions of the World," or for short, "The Book of Roger."

Mazara is "a splendid and glorious city, it lacks nothing and is without equal with regard to the magnificence and elegance of the buildings." He expounds at length about the beautiful buildings, public baths, and the gardens and orchards that embellished the city, all inherited from the Arab occupation. By then the Normans had built their cathedral, churches, monasteries, and convents around the old Arab town. Mazara remained an Arab-Norman town until the late sixteenth century, when the baroque style became the rage in Sicily, and two grandiose piazzas were superimposed on the old medieval plan to give Mazara the look of a city of wide open spaces.

While I am delivering my lecture, we drive downhill toward the sea and the city comes into view—a mass of flat-topped sand-colored buildings with a sprinkling of domes and towers. The most distinctive landmark is the church of San Francesco, with its conical bell tower clad in brilliant cobalt majolica. I was hoping to show Charles the most ostentatious Sicilian baroque interior in the town. Unfortunately we are too late and we arrive to find the doors of the church well and truly shut. Likewise the cathedral, which is closed for repairs. So it looks as if our tour of Mazara is to be one of baroque facades superimposed on medieval structures, domes and belfries, an *antipasto* instead of the true meal it should be.

As we approach the church of Santa Veneranda, dedicated to a Sicilian saint (a virgin, of course) and paradoxically a namesake of Venus, I point out the top of the church, where an eccentric architect has been allowed to have his way. The two bell towers are mounted on square balustrades, but instead of the usual twin peaks, the curved tiers of these spires taper up into graceful pagodas. Not content with the exotic shape, the architect overlaid them in local majolica in a brash checkered pattern of yellow and green. As with other Norman foundations in Mazara, the interior of Santa Veneranda was renovated and treated to a baroque stucco facelift in the seventeenth century, giving it the monochrome look of a highly decorated wedding cake. Its loveliest features are the wrought-iron balconies on each side of the nave, curving out at the front like a goose's breast and entirely enclosed by latticework. These allowed the cloistered nuns to gaze down at the Mass, at the same time making them invisible to priest and congregation alike.

This time we are able to show Charles more than the facade and the bell towers. The church doors are wide open and loud chatter from within suggests that a service has just finished. We push our way in through the crowds blocking the doorway, only to find there is a confirmation service in full swing. The priest is about to bless a pack of eight-year-olds, plump little girls decked out in white satin, gold earrings, and shiny high-heeled pumps and prepubescent boys strutting their stuff in black suits and slicked-back hair, looking more like mini-Mafiosi than confirmation candidates. As the children swarm toward the altar, failing hopelessly to affect the solemnity expected of them, their families exchange loud greetings, with the female members providing running commentaries at the tops of their voices. No wonder the priest is shouting. How else will he make himself heard over the din?

"Are you telling me this is a confirmation service?" Charles seems mildly shocked by the sheer havoc of it all, but, as I explain, this is Sicily, and Sicilians don't always feel themselves bound by rules of behavior that apply elsewhere.

Once back in the car, Marcello swings the Volvo around in a U-turn with a friendly wave to other drivers on the road, then takes us at a fast pace back to the Piazza della Repubblica. Instead of keeping to the main streets, he takes the shortcut through the Kasbah, breezily ignoring the one-way street signs. Calmly negotiating the maze of narrow lanes, he slows down only to fold back the sideview mirrors, inadvertently proving that medieval town planners were more forward-looking than one might think. They may have built their streets narrow for shade and easy barricading, but they also had the forethought to make them wide enough for the station-wagon vehicles of the future. We emerge without a scratch, much to Charles's admiration. Then Marcello reverses down a one-way street behind the cathedral and calmly maneuvers the car onto the pavement in full view of a couple of traffic wardens.

The Piazza della Repubblica, one of the most impressive baroque squares in Sicily, is viewed from the car. As we approach the promenade, I point out the tufa wall with olgive arch, all that remains of the castle built by Count Roger, Sicily's Norman conqueror. So much for the tourist's view. But while I have been earnestly pointing out Mazara's historical treasures, Marcello has been diverting Charles's attention to an alternative agenda, one much more to his own heart—the house on the seafront where he grew up, the orphanage where he cut the ribbon at the opening ceremony when he was just six years old, the *scuola media* where he had his first traumatic school experience, and the *liceo* in the ex-Jesuit college, where he submitted to the discipline of priests.

Marcello's version of Mazara is different from mine. As an outsider, I am attracted to "the relicks," the material mementos of Sicily's extraordinary history, the Greek temples, Arab kasbahs, Norman churches, and baroque palazzi. To Marcello's family all this is just the backdrop to another verità mapped out by family history. Maria, who has lived in Mazara all her life, sees her home town not as an ancient port at the heart of the Mediterranean but rather as the seat of her father's industrial empire. In the days when she was in charge of the family schedule and I had to accompany her on the daily trips to town, she would map out the city according to Luigi Vaccara's achievements.

Pointing out the *cantina sociale*, "This is where my father set up his first wine production. Eventually he bought up the whole of Via Maccagnone next to the railway station and built the family home and the factory there."

She never failed to point out the theater and the two cinemas Luigi owned, one of which has been turned into a conference center. If her mental faculties were not impaired she would have crowed with delight to turn on the television, as I did recently, to see the President of the Republic making a speech to the nation from the ex-Vaccara cinema. On our daily excursions to town, Maria would often cross the bridge over the river just for the pleasure of driving down the street named after Luigi Vaccara, reminding me again and again that the zone beyond the river, the Trasmazaro, had been a no-man's-land before her father converted it into an industrial area. We would pass the Vaccara Tonnara, where she used to watch the *mattanza*—the ritualized slaughter of tunny fish—as a child, and next to it the Vaccara Conserviera, Mazara's first canning factory. Both were closed shortly after Luigi Vaccara's death and now house a large school. There is also a sports stadium built by and named after her eldest brother, Nino, who combined intellectual interests with a passion for football. The new port with its fleet of fishing trawlers evoked more memories.

"*Tutto questo è grazie a Papà*—all this is thanks to Papà" she would say, gazing proudly at the vessels moored in the docks. "He revolutionized the fishing industry here."

Her greatest nostalgia, however, was for the *casa di Mazara*—not the grand house in Via Maccagnone where she grew up, but the house on the promenade looking out to sea that her father bought for her and Goffredo. Until the family's financial crisis forced them to sell, this vast house in a prime location between the seafront and the main square was the center of Marcello's childhood world. Now owned by a prosperous notary, the building is unprepossessing in itself. But Marcello remembers spacious, lavishly furnished rooms, the comings and goings of the extended Vaccara family, endless parties, and a bevy of servants. (The boys each had their own nanny who slept in the same room with them until they reached adolescence.) Maria was an enthusiastic hostess, and any visiting celebrities were always invited to stay. The visitors' book listed opera singers, ambassadors, politicians, artists, and entrepreneurs. In those days, before leisure time was dominated by television, the main sources of popular entertainment were opera and musical shows, and each provincial town had its own theater. When entertainers or opera stars came to perform at the Vaccara theatre, Maria and Goffredo would wine and dine them in style. One of the guests in the fifties was Claudio Villa, a great Italian heartthrob. As soon as the word got around that he was staying at the Manzo villa, a crowd gathered outside the house, clamoring to see him. When the crowd could wait no longer, Maria threw open the windows and stepped out onto the balcony, leading the singer by one hand and waving graciously with the other.

For Marcello, the house on the promenade represented the family's golden age, when his parents were young and untouched by worry. The warm home atmosphere, the ever-present brood of cousins, the freedom from the discipline of school, and the proximity of the sea all contributed to a happy childhood, and in spite of the shadow of the Mafia, a deep sense of security. Across the road from the house was one of the cinemas Maria owned. The young Marcello was obsessed with the big screen, just like the little boy in Giuseppe Tornatore's *Cinema Paradiso*. Whereas most films about the recent Sicilian past tend to focus on violence and death, Tornatore's film is unusual in that it shows what life in provincial Sicily was actually like fifty years ago. When I saw the film, I recognized the ambience of postwar Mazara immediately, and I had no difficulty imagining Marcello in the role of the little boy. The projector of the Cinema Vaccara was manned by a boy called Piero, not much older than Marcello himself. He now owns a shop in town selling electronic equipment, and whenever we drop in for a chat, he recalls how every night after dinner Marcello would turn up and beg to be allowed to help project the films. The most that Piero would allow him to do was rewind the reels.

"Do you remember, Marcellino? *Che rompipalle!* What a pest you were!"

"Why didn't you chuck me out, then?"

"Ah, because of your mother, Signora Maria. She was so kind to me. She treated me like a son; no, in fact she treated me even better than you."

It is clear that if we make a home for ourselves at Santa Maria, Marcello will simply return to his roots. But there are two of us in this venture, and as my brother has pointed out, Sicily is not everyone's idea of a peaceful retirement home. After over a decade of Mafia wars followed by the well publicized Maxi Trials of the bosses in Palermo in the 1980s, the Sicilian Mafia is well into the forefront of public attention. Although Marcello and I did not live through this brutal period at close quarters, as we followed the events from afar, we were horrified and saddened at what was happening on the island, Now, back in Italy, we are still feeling the shock waves of the 1992 assassinations of Giovanni Falcone and Paolo Borsellino, two courageous anti-Mafia magistrates who broke the back of Cosa Nostra and put Toto Riina, the "boss of the bosses" behind bars. With the pictures of the aftermath of the Capace bomb in which Falcone, his young wife, and their escort were blown to bits still fresh in our minds, only two months later we learned of the assassination of his successor. Paying a Sunday morning visit to his mother in Palermo, Borsellino was killed by a car bomb planted outside her building.

In view of the events of the last two decades and the revelation of Christian Democrat politicians acting as Mafia mediators in Rome, it is now impossible to deny the role of Mafia and how completely it contaminates the political and business communities alike. It was not always so. On my very first visit to Sicily in 1975, I discovered that, even though we were in the thick of Cosa Nostra and Maria herself had written several books about it, the subject was taboo at the dinner table. Naive as I was then, I had expected to find Mazara, if not bristling with gangsters, at least revealing some telltale signs of Mafia presence. After all, weren't we in the province of Trapani, a notorious center of Mafia activity? Wasn't Corleone a real town and only fifty miles from Mazara? And what about the regular reports of Mafia murders that appeared in the national press? When tackled directly on the subject, Goffredo and Silvio were uncomfortable and evasive. The press reports were written by northerners and were *molto esagerati.* They themselves knew very little about the Mafia. Mazara was a quiet little town. Now that the family was no longer rich, "those people" did not feature in their lives.

"*Si amazzono fra di loro*—they just kill amongst themselves," Goffredo stated emphatically. "It's none of our business." Like Goffredo, Silvio adhered to the "As long as you don't cross their path, you'll be all right" philosophy. In the midseventies, in spite of contradictory evidence, it was still possible to shrug the Mafia off as a Hollywood obsession. But their tight-lipped reluctance to open up on the subject merely confirmed my own preconceived ideas about Sicilians and *omertà*—the sacred code of silence.

Yet only one year later I was to discover that Mazara was not quite as Mafia-free as my hosts had made out. By this time Marcello and I were officially *fidanzati*, and he had brought me back to Sicily as his bride-to-be. We had just landed at Punta Raisi, as the airport of Palermo was then called, before it was renamed Falcone and Borsellino. Silvio picked us up in a borrowed car and accompanied us back to Mazara. We were circling Piazza Mokarta, Mazara's main square, when we were accosted by one of the many Vaccara cousins who go by the name of Luigi. Very much out of breath, he was extremely excited about something.

"*Hanno amazzato uno. Venite a vedere!*—They've just killed someone. Come and see!"

As *The Godfather II* was still doing the rounds of Italian cinemas, at first I assumed I was the victim of some kind of practical joke. It turned out this was no joke. When we arrived at the garage on the outskirts of the town, we were told a Mafia murder had just taken place. By the time we got there, there was nothing to see but three bullet holes in a wall. A garage hand had been shot dead, the body had been removed, and the Carabinieri were taking measurements. There were no eyewitnesses. If Luigi was disappointed to have lost a scoop, I was incredulous. Did this kind of thing take place on a regular basis? What I found most disconcerting was that I was the only one to be shocked by the event. Why had this happened? Did anyone know? With a slight shrug of his shoulders, Silvio murmured *"Cose loro."* He was echoing Goffredo—"other people's business." (Other people's business turned out to be a decade-long vendetta conducted between a Mafia boss and his brother-in-law in a nearby town, with ramifications all over the province of Trapani.) Marcello and Silvio seemed not so much shocked as embarrassed—embarrassed that an outsider, and especially a foreigner, should get a bad impression. Having been in apparent denial until then, they had to admit that this kind of incident did occur, but only very rarely. I was dismayed, not just at the murder, but at their phlegmatic reaction. As yet I had not been initiated into the Sicilian obsession with face saving, and I did not realize that they were playing the episode down for my sake. Appearances were important, and above all it was important to protect a guest from unpalatable truths. This explained their embarrassment at the very mention of the Mafia and now their reluctance to admit its pervasiveness. By offering their bland version of la verità, they were maintaining the highest standard of Sicilian hospitality.

Once I had become an official member of the Family, they lowered their guard, and I learned what I had always suspected—that both the Manzo and Vaccara families had had their brushes with the Mafia. Maria was the least evasive on this subject, and so it was to her that I put the question: How did Luigi Vaccara manage to build up an industrial empire without cooperating with the Mafia? Surely he must have been an obvious target for intimidation? Maria had regaled me with stories about boarding school in Florence, where she was sent at the age of six. What she hadn't told me at first was that she and her brothers were sent so far away from Sicily because her father had received extortion threats and feared his children might be kidnapped. For the same reason, Marcello and his brothers were educated at home with a governess, not attending local schools until they were in their teens. Any family of wealth and standing in those days were targets for Mafia kidnappers. In answer to my question, Maria simply said that her father knew how to handle the situation. Besides, he had built up his industrial empire in the twenties and thirties when the Fascists had the Mafia under control to some extent. Things got much worse after the war, Maria said, but her father was dead by then. Her own generation had had a far more difficult time, that is, as long as they had commercial and industrial interests. She herself had had no contact, for although she was the property owner in the family, the Mafia would not do business with a woman. On one occasion, she recalled that Goffredo had put in a bid for a piece of land to plant a vineyard when he received a note advising him that the land was no good and it was not a good idea for Goffredo to buy it.

"Was that all?"

"That was all."

"And did Goffredo buy it? Did he go to the police?"

"Of course not. *Sei matta*—Are you mad? Anyone who does business here in Sicily has to live with those people, just as we had to live through Fascism." Maria, whose optimism normally knew no bounds, never expressed any hope that Sicily could be rid of the Mafia.

In the 1930s, thanks to the exertions of Mussolini's iron-fisted prefect, Cesare Mori, the Mafia was driven underground, but with the fall of Mussolini's government, it made a vigorous comeback. In preparation for the Allied invasion of Sicily, the American government released Lucky Luciano, the head of the U.S. Mafia, from a fifty-year prison sentence. His mission was to escort the American troops and make the necessary "contacts" who would welcome the invaders and organize provisions for them. With the American forces unwittingly reinstating the Mafia and the Christian Democrat government actively colluding with it, the organization mushroomed after the war with tragic consequences for Sicily. Hence, in the late forties and early fifties Goffredo and the Vaccara uncles were once more reduced to paying the *pizzo*—protection money—not only to traditional Mafia henchmen but also to Salvatore Giuliano, the notorious robber king of western Sicily. Giuliano had written to Goffredo complaining about a farm manager who worked for the Manzos and the Vaccaras. This man handled the regular payments that both families were obliged to make to Giuliano, but apparently he was not treating the bandit with due respect. Giuliano threatened Goffredo and Maria's brother Nino with reprisals if they did not immediately get rid of the man whom he described as a "*verme della terra*—worm of the earth," a phrase that shows the bandit had a certain way with words. This letter in Giuliano's crude handwriting with its atrocious spelling (according to Goffredo) was in the family's possession until it was sold to pay off a debt. Only months after his threat to Goffredo, Giuliano was murdered in his sleep in nearby Castelvetrano, probably by his cousin Pisciotta (who was poisoned in his turn in the Ucciardone Prison in Palermo).

Not everyone in the Family was prepared to toe the line to the Dons and the bandits. In the early sixties a Manzo cousin dared to defy the Mafia and fell victim to them accordingly. A brave and independent man, he refused to pay the pizzo on his meat-processing business. Inexplicably he agreed to meet the men who were threatening him and got into their car, only to be stabbed to death from the back seat. Before he left for the fatal rendezvous, however, he had had the foresight to leave a letter for his family with the details of where he had gone and exactly who he was going to meet. As a result, the thugs who murdered him were jailed for life at a time when Mafia crimes were rarely punished.

To my surprise there still seems to be a vestige of nostalgia about the Mafia of Maria's youth, when the "Dons" were known as "men of honor" and well-known Mafia chieftains, the likes of Don Calogero Vizzini, had infiltrated the establishment to become mayors and members of parliament. Zu (Uncle) Caló, as he was affectionately known, was officially nothing more than the mayor of Villalba, a small hilltown community, but thanks to the American intervention, he had been the key man in the Mafia's resurgence as a political force. In those days an aura of glamour still adhered to such men and the distinction between 'men of honor' and gentlemen was often blurred. Even for Sicilians it wasn't always easy to tell who was a Mafioso and who was not.

Maria was once visited by a distant cousin from Canada, who admired her art collection, including a painting she claimed was by Correggio. Realizing that the family was in financial straits, he made an offer they could not refuse. His purchases had already been shipped to Canada, and he had even offered Pilli "work experience" under his tutelage, when Maria and Goffredo discovered the true identity of their dinner guest: he turned out to be a well known member of the Mafia in Canada.

Even now, Mafia connections are wont to turn up in unexpected places. At dinner one night Silvio's wife, Nanette, announces that the son of one of the Vaccara cousins is dating a girl whose father has been arrested for strangling seven men "with his own bare hands." The name of the killer seems to ring a bell for Marcello.

"What was his name again?"

"You must remember him. He was in your biology class."

After graduating from university, Marcello did a brief stint as a teacher at the *liceo*. He now remembers the man in question as an acned, unprepossessing youth.

"Did you say with his bare hands?"

This conversation is moving into the bizarre. We could be characters in a Grimms' fairy tale marveling at the man who boasted "seven in one blow," but in the fairy tale the victims turned out to be flies, not humans. If a Vaccara is dating the daughter of a Mafia killer and the relationship blossoms, where does that leave us? Will we be invited to a Mafia wedding? Will we actually have to shake those strong bare hands? This is indeed a sobering thought.

Even more sobering is the extent to which the crackdown on the Mafia gives the state the power to infringe the civil liberties of the ordinary citizen. Some years ago a close friend was charged with "association with the Mafia." He then spent a horrific three and a half years without trial in the Ucciardone, the notorious jail fortress in Palermo. The evidence against him was based on the statement of a *pentito*, one of the controversial Mafia informers responsible for the capture of Toto Riina and the prosecution of the veteran politician Giulio Andreotti. Just for good measure they arrested his brother too and kept him in jail for a year, but were forced to release him when the pentito withdrew his statement. When our friend's trial eventually took place, he was sentenced to three years and five months. As the sentence coincided with the time he had already served, conveniently he could be released without loss of judicial face. On appeal the sentence was reversed, and he was finally proved innocent. In the meantime his life, his family, and his business had been all but destroyed.

In the wake of the Falcone and Borsellino assassinations, there has been a universal repudiation of the Mafia, but there is a feeling in local circles that certain magistrates engaged in the anti-Mafia crackdown are abusing their powers for political or personal ends. The latest victim of an overweening magistrate concerns a man we know who owns a factory in a neighboring town. In no way involved with the Mafia, somehow he became embroiled in a personal feud with a vindictive branch of the law. The origin of the dispute, connected with some bureaucratic regulation he had flouted, is now lost in history. To make his point the magistrate sequestered the factory, encircling it with police so that no one could go in or out and production had to stop. However, the magistrate overstepped his powers by having the owner's phone bugged. Well aware that he was being overheard, the man phoned a friend denouncing the magistrate as a cuckold. *Cornuto*—cuckold—being the worst possible insult anyone can make to the Sicilian male, the magistrate was furious and suspected that he had been disgraced by none other than his worst enemy. But at the same time he was helpless to confront the man whose phone he had bugged without authorization. His wife was indeed pregnant, and as soon as the baby was born, a DNA test confirmed his worst fears. He, the magistrate, was not the father, but neither was the industrialist, who had in fact never even set eyes on the woman. Meanwhile everyone but the elderly magistrate had known that his pretty young wife had been busy "putting the horns on him." Naturally, the outcome was a humiliatingly public divorce.

The myth of the Mafia as a product of Sicilian passion and pride, a code of honor adhered to by a backward people, nothing more than a perverted form of rustic chivalry, is now well and truly obsolete. There is no doubt that the Mafia is a highly organized secret fraternity that pursues power and money by intimidating and killing people. Yet, this being Sicily, nothing is black and white, and a Pirandellian element of relativism still clouds the picture. Some Sicilians, like my brother-in-law Pilli, insist that Cosa Nostra's fame has exploded as a result of international media exploitation. According to Pilli's *verità*, the Mafia exists, but it is replicated everywhere and the Sicilian version is no worse than any other. The same thing goes on in the United States, in Canada, and even in Britain, he claims. You have to be brave to take on Pilli in an argument, for he has what is known in the family as "the Manzo Temper," which means that a mere difference of opinion can result in a volcanic eruption of wrath. Yet in the safety of my study in Milan, well away from the shouting and table thumping of my brother-in-law, I can have my say. Surely Pilli is wrong and the Sicilian Mafia is different. It is the ruthless capacity to kill that distinguishes Cosa Nostra from other criminal organizations. Corruption and protection rackets exist the world over, but elsewhere the perpetrators of organized crime do not dissolve children in acid baths, carry out mass murders, or blow up magistrates.

To anyone interested in Sicily and the Mafia, the question remains: what is so different about Sicilians that such a ruthless organization could flourish here for so long? History has as much to do with it as Sicilian character. The Mafia is a relatively recent phenomenon which emerged as an organization a century and a half ago in the political and social chaos that followed the Unification of Italy. Unofficially its roots went way back before that. Whatever its origins—and there are good reasons to think that these may date back to the Roman system of *clientela*—patronage—social conditions on the island provided fertile ground. For hundreds of years Sicily was a feudal society ruled by foreigners. Most Sicilians worked on the land and owed their living to an absentee baron who administered his property through *gabellotti*—henchmen—who, more often than not, abused their power to intimidate and extort from the backward *contadino*. These men were the forerunners of the Mafia of today, whose power base has shifted away from the land to control of the drug market and the even more lucrative arena of public tenders. All this may be so, but similar conditions prevailed elsewhere in the Mediterranean. How do we explain Sicilian society's acceptance of such an organization? Is there such a thing as Sicilian character, and if so, is it responsible for acquiescence with the Mafia?

Leonardo Sciascia, one of the greatest Italian writers of the twentieth century, was the first Sicilian to break omertà in print and openly challenge the Mafia. Whereas his mentor, Pirandello, had treated verità as relative, dwelling on the impossibility of arriving at an objective truth, Sciascia was not content with that. His Sicily was not just a world of unreason and psychosis. In *The Day of the Owl*, published in 1961, he explicitly denounced the Mafia, while his contemporaries were still donning Pirandellian masks in an attempt to conceal the phenomenon. His first and most famous novel revolves around a murder and cover-up of truth in a small town in Sicily. No ordinary crime story, it breaks away from the traditional pattern in which a professional investigator manages to identify the murderer, and the guilty party meets due punishment. In this case the hero is a low-ranking policeman, a quixotic character who does not share the community's fear of the Mafia and resists pressure by insistently pursuing his investigation. Meanwhile everyone else already knows exactly who is responsible for the murder and helps to shield him. The little man, the antihero-cum-heretic, pursues a truth that the majority of characters are bent on twisting and obscuring. *Omertà* is brutally exposed as part of the ethos by which the truth is turned inside out and falsehoods are created which everyone pretends to believe.

Sciascia's early books deal with the local rural Mafia, but his writing then moves up the echelons of society and out of Sicily to expose corruption in the boardroom and in government, and the power of the Mafia at the heart of Italian democracy. His historical novels make the point that the eternal enemy is not any particular individual, but the system itself. The state is weak, and those who have jostled to the top, be they inquisitors, government officials, priests, or Mafia dons, are driven by an ever-increasing greed for power, while the people, sedated by fear and apathy, stand passively by. According to a Sicilian proverb, "*Cumannari è megghiu chi futtiri*—wielding power is better than screwing." Power corrupts and its victims are corrupted in turn. Only in this context is it possible to understand the Sicilian subjection to Cosa Nostra.

When Goethe wrote that Sicily was the key to Italy, he could hardly have been thinking of the Mafia, for the phenomenon did not even exist in his day, but his precept is equally valid for the network that Italians also call *la piovra*—the octopus, whose tentacles have reached the very heart of government in Rome. Sciascia was certainly thinking of the Mafia when he wrote, "All my books in fact are a single book: a book about Sicily, which touches the sore points of the past and the present." But his books are not just about Sicily or even Italy. As he said himself, "I have used Sicily as a metaphor, let's say for the world. The evils of the microcosm of Sicily are the evils that can bring death to the whole world."

Passionate as I am about Sicily, and optimistic about the chances of driving the Mafia to ground, I have to admit that there are certain aspects in the culture that make it easier for the Mafia to survive. First and foremost is the sense of duty to the family. When an individual has to decide between obligation to the state or to the family, his conditioning will prompt him to choose the latter. In this respect nothing has changed in Mediterranean society since Sophocles' Antigone defied authority to bury her brother on pain of her own death.

Secondly, intrinsic to the Sicilian way of life is the tradition of the mutual exchange of favors. "One hand washes the other," as Maria used to say. The natural consequence of these two traits in the Sicilian psyche is respect for the local power-wielder-cum-benefactor and contempt for the "ball-breaking" authorities, an attitude that the Mafia is only too glad to exploit. Some of the dubious forms of influence they adopt are not only common, albeit in a diluted form, but even intrinsic to the Sicilian way of life. Intimidation of the kind Goffredo was subjected to in the 1950s still goes on. Local politics are still influenced by the Mafia, and any candidate for the position of *sindaco*—mayor—has to reckon with this fact. In a recent local election a young man without political experience campaigned on a Clean Hands platform, promising to fight the endemic Mafia pressure in the allocation of government tenders. Adopted as the young people's candidate, he was elected by a large majority, only to stand down just days after his election, pleading that he was "not well enough" to take on the responsibility. The reason for his about-face was obvious and a huge disappointment to all those who, in voting for him, had hoped that the Mafia tide was about to be turned at last. The sindaco elected in his place had no such anti-Mafia reputation, but even he came up against intimidation tactics. When an unpopular new street parking proposal was due to be discussed at the comune, the mayor found that his car tires had been slashed. Shortly afterwards the plan was dropped.

Intimidation is an occasional hazard but far more pervasive is *racco-mandazione*—string pulling, not just an expedient in exceptional circumstances, but a normal way of getting things done. Again, Pirandellian relativism sets in here whenever this subject is debated at Santa Maria. As I see it, there is nothing wrong with "recommending" a friend to the owner of a restaurant or a hairdresser, but when it comes to help someone pass an exam, get a fine waived, or secure a job, it begins to look like cheating. Is this practice not just a shortcut that confounds the system and takes unfair advantage of others? Pilli and Maria disagree with my simplistic black-and-white approach. According to their *verità*, everyone has strings to pull and everyone pulls them. If they didn't use raccomandazione to help the family along, they would be neglecting their duty. Maria was the most insistent and systematic practicer of raccomandazione I have ever come across, and her prowess in this field largely accounts for her success in saving Santa Maria. It was a vital survival skill, which she deployed regularly and shamelessly in the interests of the Family. On one occasion she had just put down the telephone after a long conversation with a well-known member of the Sicilian senate.

"That was Senatore Vanella on the phone," she proclaimed triumphantly to a crowded room. "He was so kind and helpful. Now does anyone need a raccomandazione?"

Yet to be an agent of raccomandazione doesn't feel quite so comfortable. One of Marcello's cousins is a teacher at a local secondary school. Every year she deals with parents asking her to doctor exam results so their children can be *promossi*—passed up to the next class, and, strict professional that she is, every year she has the unenviable task of disappointing friends and friends of friends, inevitably causing offense. Silvio sits on selection committees in his role at the Soprintendenza, and it is not unusual for him to receive calls from the relatives of candidates, attempting to influence his vote. It would be rude not to lend an ear, but he does no more than that. When the decision goes against the "recommended" candidate and he has to justify the committee's choice, he goes into paroxysms of embarrassment.

While cultural attitudes are rooted in history and that history still rests heavily on Sicilian shoulders, there are signs that things are changing. If Goffredo was subjected to *intimidazione* and Maria relied on raccomandazione, their grandchildren or at least their great-grandchildren, will inherit a different Sicily. The passivity that supports the Mafia is being gradually phased out as submission gives way to rebellion and outrage. The Mafia still exists, but the successes of Falcone and Borsellino and their colleagues in identifying and combating key members of the organization have given Sicilians a hope they had never had before. Toto Riina, the boss of Corleone, has been put behind bars for thirty years, and Bernardo Provenzano, his henchman and heir, has been captured by the Italian police after forty-three years on the run. The Sicilian people, who reacted with such desperation and horror to the assassinations of their heroic magistrates, are turning their backs on omertà and speaking out. The war is on. No one is making excuses for the Mafia any more. And thinking of my brother's misgivings about making a home in Sicily, I say, yes Charles, I am hopeful too. No way am I going to stay away from Sicily, just because of Cosa Nostra.

FOUR

The Cascina Fiasco

CHARLES DIDN'T STAY LONG IN SICILY. WITHIN DAYS HE DE-
parted for the comfort of his nicely ordered, caper-free air-conditioned
mansion in Bangkok. As he waved good-bye at Palermo airport, he
couldn't resist a parting shot, "Remember the Cascina Fiasco." He need
not have reminded us, for it was with the Cascina Fiasco in mind that
we had approached the Santa Maria project with so much caution in
the first place. The Cascina Fiasco took place on our return to Italy after
a series of jobs in Latin America and the Far East. Expatriate life means
rented accommodation, and by then we were thoroughly sick of it. We
had lived in other people's houses for so long that the prospect of
repeating the experience in Milan was extremely unattractive. When I
say *we* decided, the first-person plural has a touch of the royal, for I have
to admit that, desperate for a home of our own, I was the driving force
behind the venture. Unlike Marcello, I had never had a family home to
link to memories of childhood. Whereas he had always had the security
of Santa Maria, my childhood had been a kaleidoscope of shifting
images, a saga of moves between rackety bungalows in Fiji, where my
father was a district commissioner, to an assortment of London flats,

where we camped when the family came home on leave, courtesy of an eccentric Irish grandmother. Life with Marcello had followed a similar pattern. We had chalked up a bizarre collection of homes, moving house roughly every three years. The return to Italy was to herald a new permanence in our lives. It was time to put down roots and buy a house. It didn't take me long to find one. My choice was a *cascina*, one of the traditional farmhouses of Lombardy and Piedmont in the flat-lands of the Po Valley, fifty miles from Milan. It was built around a courtyard like a Roman villa, designed to house not just the landlord and his family but a whole farming community, including laborers' houses, stables, and barns. It also had a chapel with a belfry, a mature garden, and a lovely brick gazebo. Having been abandoned for almost thirty years, it was in very poor condition. All the better, as far as I was concerned. I wanted a renovation project I could get my teeth into, and besides, it had so much going for it. The main house had a marble stair-case, frescoed ceilings, spacious rooms, terra-cotta floors pitted with age, and a huge fireplace in the kitchen. Included in the complex of buildings was the shell of an eighteenth-century house that had never been renovated. From the cracks in the walls I pulled out great swabs of horsehair, an original component of plaster from the Age of Reason. It was all too wonderful. I persuaded Marcello that we just had to have it.

Meanwhile our friends in Milan were alarmed at our decision to forgo civilized life in the city for the outback of rural Piedmont. They told us we couldn't go and live on farmland amongst the contadini— we'd hate it. Beware of the wily Piedmontesi, they warned us. They'll eat you for breakfast. The location was wrong, wrong, wrong. There was no view, the fog in winter turned the roads into death traps, and the heat in summer was something terrible. Carried away by the dream of owning a villa in the Italian countryside, we ignored this well-meant advice and went right ahead with the madcap venture.

As thoroughly wet behind the ears, first-time buyers of Italian real estate, we approached the purchase of the cascina with an enthusiasm that must have kept the locals chuckling into their glasses of Barbera for

months on end. Finding the asking price most reasonable, we duly
handed over a hefty sum for the *compromesso*—deposit, no questions
asked. After obtaining a survey and an architect's estimate on the work to
be done, we took out a mortgage to pay for the renovation work. Then
we rented a house in the village in order to be close to the cascina. No
sooner had we moved in, than the problems began to crop up. Connecting to the electricity and water supply had not been budgeted for. Nor
had replacement of the stable roof, which began to sag dangerously as
soon as we handed over the money. When the builders' estimates came
in, they were two to three times higher than the architect had foreseen,
propelling our budget further and further out of control. Meanwhile,
installed in our rented house in a one-horse town in Piedmont, we were
getting a taste of the mosquito season in the Po valley and not taking to
it one little bit. The great whining creatures issued forth from the neighboring rice fields like tiny airplanes to torment us day and night. All this
was bad enough, but then one day by chance I discovered that the elderly and *simpatico contadino* who had sold us the cascina, had at the same
time sold the adjoining field to a gravel mining company. How could
that be? How could he have been so treacherous? Or rather, how could
we have been so gullible? There must have been some mistake. I checked
with the comune. The town surveyor was surprised we didn't know
about it. Everybody else in the village did. But being tight-lipped
Piemontesi, they had opted not to pass the information on to us, the
people concerned. He confirmed that the mining company had been
granted permission to excavate the field for six years. What? Six years of
drilling right next to our dream-home-to-be? Bulldozers thundering up
and down the country lane? And after six years what would we be left
with for a view? A whopping great pit. There had to be a law against it.
Apparently there wasn't, at least not at that time. Not one of the regis-
tered letters I sent to local and regional government officials received an
answer. Without actually saying I told you so, Marcello had already come
to the only sensible conclusion—that the whole venture had been a terrible mistake. Fortunately we were in time to cancel the mortgage. But

we were still left with the great white elephant of the cascina. We just couldn't wait to get rid of it.

Getting rid of it wasn't going to be as easy as buying it had been. By now our eyes had been peeled open to the fact that we had paid far too much for it. When we bought it, we did not realize that outside Tuscany and Umbria, where they can be made into commercial enterprises, here, oversized derelict buildings in the middle of nowhere, no matter how historic, are practically given away. Nobody wants to live in them, and the cost of renovation is guaranteed to escalate to such an extent that it is far cheaper to build from scratch. There was the additional problem that in the sale contract we refused to sign to a value lower than the amount we actually paid. This is common practice in Italy but can cause problems, when you want to sell.

"You can't insist on declaring the real value," we were told at the time. "You'll be like 'white flies'—loathsome outcasts." For months we advertised it for sale in upmarket property magazines without success. Meanwhile there was a devastating flood in the area and our cascina, though undamaged, suddenly plummeted in value because it happened to be in the flood zone. It looked as though we were saddled with it forever, gravel pit and all, when a friend put us in touch with the owner of a construction company. His wife had just died of cancer, but she had been very close to a certain priest-cum-television-personality, regularly appearing on one of those interminable variety shows that dominate Italian television on a Sunday. As a tribute to his wife, the building magnate wanted to acquire an isolated country property to donate to the priest, who was looking for a site for a drug rehabilitation center. Since it was a long-term project, the imminent gravel works did not put him off, and to our great relief, he bought it.

We had escaped financial ruin by the skin of our teeth. Although my derring-do venture had put marital relations under considerable strain, we survived. Even so, Marcello was adamant. Having lost all faith in my nose for the property market, he made me swear on pain of divorce never, ever to suggest buying property in Italy again.

"See what happens when you get involved with northerners?"

I had seen. There was no denying that we had plunged way out of our depth into what turned out to be a sea of sharks. I duly swore we would take no more risks from then on.

Some years have gone by since the debacle of our one and only enterprise in Italian real estate, but we didn't need Charles's reminder to approach the renovation of Santa Maria with caution This time, however, since Marcello already owns his share of the property, there is no purchase involved, no throwing ourselves at the mercy of ruthless estate agents or wily contadini in sheep's clothing. On the other hand we are once again contemplating the renovation of an old house in Italy, and this time at long distance.

On reflection, there's something else that Marcello and I are going to have to thrash out, something that has been bothering me for some time—the prospect of communal life at Santa Maria. Privacy (and Italians don't even have their own word for it) is very important to me. An Englishwoman's home is her castle, as my version of the saying goes, but in this case the castle will be shared between three families and no doubt countless Vaccaras and Manzos will be afforded ready access. If we make Santa Maria our home and eventually retire here, what happens to the privacy that the Anglo-Saxon tradition holds so dear? Marcello deals with this objection, pacing the room and stroking his mustache.

"Hey, do you want to be *solo come un cane*—lonely as a dog? *Per me la privacy non è un problèma.* If you want to be *isolata*, you can have your own study and lock the door."

An Englishwoman's privacy is a Sicilian man's isolation. By the Goethean principle Italian sociability is taken to an extreme in Sicily, and privacy, or "isolation," as Marcello calls it, does not come high up on his list of priorities. Maria's gregariousness, according to which, time spent *in compagnia* is infinitely preferable to time spent alone, runs thick in the veins of all three of her sons. And with three families living together, the chances of any of us spending time as lonely as a dog are exceedingly slim. Besides, as Marcello points out, the advantages of communal living are manifold: if we run out of coffee or milk, if the car breaks down, if the phone is cut off, if we need inspiration for a recipe, for any emergency in fact, all we'll have to do is shout to Silvio and Nanette Upstairs or Pilli and Maria Antonietta Next Door for supplies or assistance. What could be more convenient than that? And best of all, we'll never be short of compagnia.

"You got your way over the cascina, didn't you? And look what trouble *that* landed us in. Enough of faintheartedness, Caro*line*. (After twenty years of marriage he *still* says my name with the accent on the wrong syllable.) Come on! *Coraggio!* Santa Maria can wait no longer. This time we'll go for it."

What can I say? It's not as though I have a brilliant record in real estate investment. With Marcello's exhortation, the issue seems to be settled. The problem now is where to start? With the gardens, or repairs and renovation of the ground floor?

"Casa fatta e vigna sfatta," says Marcello, recalling a Sicilian proverb that means first things first: see to your house before you worry about the vineyard.

Silvio and Nanette, who have been patiently waiting in the wings all along, are relieved to know we will share the burden of expenses of this *castello in rovina*, pleased that we will now be coming every year after all. In the years since the idea of the renovation was first mooted, Silvio has been making plans on our behalf. Tucked away in the labyrinth of his study is a bundle of handwritten sheets covered with incomprehensible calculations in cubic meters. Once again he pulls them out. They are as baffling as ever, but now that we have finally decided to go ahead, we pore over them with renewed interest. What we can't understand right now is how much it will cost. Silvio points out that his calculations have not been updated and do not include electrical work or plumbing. Yes, Marcello insists, but roughly, how much will it cost? There doesn't seem to be a straight answer to this question. Silvio is an architect specialized in restoring churches. Most of the buildings he works on are hundreds of years old. He warns us that you cannot forecast the cost of a restoring this kind of building, just as you cannot predict what flaws or complications you may come up against. The local building firms may give you an estimate, but there will always be the loophole clause to cover anything unexpected. When things get sticky, they charge you for their time and then all your estimates go up the spout. Marcello in his capacity as an executive in an American company does not like this line of talk. He tries to nail his

brother down, to extract some kind of bottom line, but Silvio is expert at the art of noncommitting. Words die away. With a downturn of the mustache, a slight tilt of the chin, or a turning up of palms he expresses bewilderment, perplexity, an inability to answer the key question through no fault of his own of course.

We walk through the ground floor rooms, racking our collective brains for ways of bringing down the costs.

"What if we did the project in two stages?"

"But you've got bedrooms and bathrooms at one end and the kitchen and living rooms at the other. Which half do you want to do first?"

"We could start with the bedrooms and use the little kitchen where Maria used to cook for the dogs. That way we could leave the big kitchen and the rest till later"

"The dog kitchen is the only place the main bathroom could go."

"Well, couldn't we do up just the mini-flat we have planned for the two girls?"

"There is only one tiny bedroom in there. Where are you all going to sleep? In any case, you are going to have to do something about the damp in the rest of the house."

In short, there is only one feasible project, the one Silvio is holding rolled up under his arm. My heart sinks as we make our way through the living area. What was once a *grande salone* with its graceful, vaulted ceilings and connecting archways is now a scene of neglect worthy of a Mediterranean Miss Havisham, with flaking plaster, rat droppings, and broken window panes hanging from rotting frames. The original kitchen, which once had a tiled wood stove extending from wall to wall, is now just a shell. The stove and all the fittings and plumbing were ripped out after the 1968 earthquake, when the wiring went up in flames. This means that as far as the kitchen is concerned, we will have to start from scratch. As for bathrooms, the only one on the ground floor is a tiny shower room tucked into the tower at the other end of the house, last used by Pilli's family twenty years ago. On my insistence, we decide to go and have a look. The door is wedged shut, and Silvio isn't keen on forcing the issue. He has a rodent phobia and he knows that the open slits in the tower provide access to nesting rats, which have also taken over most of the light fittings. Marcello and I are used to rats after years of cohabitation in Jakarta, listening to them scuttling on the tin roof at night, watching them creep into the garden along the open drains. We decide to risk it. We push, and the door gives way. The corpse of an overweight rat, whiskered and heavy-tailed, plops right onto Silvio's foot. As we stand in the vaulted room with a blinding sunset streaming in through the barred window, there could be no more decisive confirmation of our decision than this putrid bundle at our feet. If we hadn't decided already, it would have been the rats to push us into going ahead. Who will be the ones to make a home at Santa Maria—them or us?

MARIA AND GOFFREDO

FIVE

Getting Close

"ARE YOU ALL AT HOME? GOOD. *AVVICINIAMOCI*."

Giusi, the youngest of Marcello's Vaccara cousins, has just announced her intention to Get Close—the expression Sicilians use for "dropping in." She comes bearing invitations to her daughter Vanessa's forthcoming wedding. A wedding! We are all thrilled at the prospect. In all these years of coming to Mazara our visits have never coincided with a Family celebration. And with three hundred guests invited to a reception at Giusi and Gianni's villa by the sea, this promises to be a humdinger of a Sicilian wedding. Still, we have been taken by surprise, partly because we had no idea that Vanessa was even engaged, let alone about to be married. In this part of the world marriage is usually preceded by a long, drawn-out engagement. Goffredino, Pilli's eldest son, has been *fidanzato* to the same girl for fifteen years, and they still haven't set a wedding date.

"Congratulations Vanessa! Who's the lucky man?" Not a Mazarese, surely, or we would already have known about this. Giusi doesn't look too pleased at her daughter's heretical choice of a partner outside the Family orbit. Glumly, she announces, "No, Vanessa has chosen a *forestiero*."

The word makes no distinction between the foreign and the alien, lumping foreigners and strangers in one single undesirable category.

Vanessa giggles, "*Forestiero!* But Dario comes from Palermo. You remember Dario, don't you, Car*oline?*"

Racking my brains, I recall a party the previous summer when we met a tall blond ponytailed young man strumming a guitar at a poolside sing-along. I had then mistakenly assumed he was just one of Vanessa's many admirers. He seemed nice enough—cool according to Clem. Surely he doesn't deserve to be treated like some kind of Trojan horse with insidious intent on the Vaccara Family citadel.

Palermo is less than a hundred miles from Mazara, so if a Palermitano rates as a forestiero, then what exactly was I, when Marcello presented me to the family as his fidanzata all those years ago? A much quoted proverb in this part of the world is "*Moglie e buoi dei paesi tuoi—*wives and oxen are better from home." Was that at the back of the family's mind, I wondered at the time, as they made their appraisals? Were Marcello's relatives shaking their heads behind our backs and making bets on how long our relationship would last? I had to admit, from the Sicilian point of view, the omens were not good. In a society where sensible people prefer their livestock and life partners to be raised locally, Marcello had taken a huge risk in choosing a foreign wife. We were of different nationalities and different religious denominations. Indeed, my being Anglican aroused much curiosity in a family of Roman Catholics. (Don't you believe in the Virgin Mary? What, not even the Pope?) Morever, Marcello spoke very little English, and my command of Italian was still somewhat shaky. How were we going to communicate?

What with Marcello working in Central America and me commuting between London and Milan, the time we had managed to spend in each other's company had come to a grand total of seven and a half weeks. The decision to get married on the basis of such a brief acquaintance must have raised a bushy eyebrow or two. Not that we boasted of it, and even Maria had the good sense to keep quiet on this point. What could not be hidden, however, was my lack of roots, which must have seemed particularly suspect to the traditional family I was marrying into. They tended to attack the subject sideways, with circuitous questions about my parents, my brother and sisters, and relatives in general. Where do your parents live? How long have they been living there? And what about your grandmother? Does she live with them? No? What? You mean she lives alone? Under this intensive cross-examination, the truth had to and did eventually out: both my mother's and my father's kin were sadly rootless. Carelessly they had let themselves be scattered around the globe. My mother was born on a sheep farm in New South Wales, my father in an obscure farming community in New Brunswick called Macnaquac, to which neither he nor anyone else in the family ever returned. This promiscuity of genesis the Sicilian Family was prepared to forgive. Perhaps the unconventional circumstances of my parents' birth had been aberrance on my grandparents' part. And before that? I hated to disappoint them, but my grandparents also had what appeared to Sicilians as outlandish backgrounds, ranging from the Antipodes to colonial India. Only one grandparent was actually born in England.

"Well, you grew up in England, didn't you?"

"Actually, I grew up in Fiji."

"Fiji? Where on earth is that? *Matruzza mia! Povera* Caroline. What a strange family!"

According to a Sicilian proverb, "*Cu nesci, arriniesci*"—you have to get out of Sicily to succeed, but this fundamentally applies to contadini. For the well-to-do Sicilian, emigration is an alien idea, and it was hard to explain the whys and wherefores of the Anglo-Saxon diaspora to my prospective family. It soon became clear just how firm Marcello's roots were compared with mine. His parents, grandparents, great-grandparents, and generations of ancestors had sprung from the soil of western Sicily. The Vaccaras had been sharecroppers, shepherds, or cowherds, as the name suggests (*vacca*—cow), until Maria's father achieved his meteoric rise to wealth. The Manzos, originally from Spain, had been public notaries in the nearby city of Trapani since the seventeenth century. Theirs was a stable and rooted family; mine was not. Fortunately Maria and Goffredo, being exceptionally broadminded, were ready to wave aside my lack of credentials as a trivial consideration. Maria, in her usual Pollyanna mode, assured her relatives that her future daughter-in-law came from a *buona famiglia*, patting my arm as she did so. I could tell that not all the Aunts were entirely convinced, for after all, what was a young girl doing living on her own in Milan so far from home, if she really did come from a "good family"? Changing tack, Maria would insist that they had to admit that her future daughter-in-law, the *inglesina*, was *carina*—pretty—while Goffredo, ever the gentleman, was gallant enough to suggest that I could pass for an American actress—American, not English, mind you, because Englishwomen, according to Goffredo, were supposed to be blonde. Finally, as a gesture of goodwill to the potential buyer, Maria came up with an add-on: Marcellino's *fidanzata* had brains and an Oxford degree to prove it.

In a way I was something of a prize, my Englishness bestowing kudos I had done nothing to deserve. I knew they meant well, but even so, I felt like a cow at a cattle market. I was looked up and down, stroked, patted, and assessed—first for Breeding, then for Beauty, with Brains thrown in as an extra. Fortunately, Marcello was more sanguine than I was. He had inherited enough of his mother's sunny outlook to dismiss my nervous misgivings with a blithe "*Non preoccuparti*—don't worry." Only later, when I got to know the family better and understood the way their minds worked, did I realize that he was right. Not that my worries had been unfounded. The fact was, I had been welcomed into the Family simply because they trusted one of their own. They may have had doubts initially and there may have been a certain amount of speculative gossip, but if Marcellino, by common consent *bravo e onesto*—a good honest boy—had made his choice, then general approval would follow.

VANESSA AND DARIO'S WEDDING IS TO BE HELD IN THE TINY church of Santa Caterina, one of the baroque jewels of Mazara. It takes place at the tail end of August, when many of the summer visitors have already departed. Much to her chagrin, Melissa has to leave for a previous invitation. She's been waiting all her life to go to a Family Wedding, and now that the chance has come, she's going to miss it.

Clem and I, both in little black dresses, cocktail party attire being the local dress code for weddings, and Marcello on sufferance in suit and tie, arrive in plenty of time to find seats near the front of the church. The ceremony, scheduled for the late afternoon, finds the guests still wilting in the afternoon heat. The ladies, most of whom, like us, are dressed in black, but unlike us, are dripping with gold and pearls, have had the foresight to bring fans. Not that the fans are particularly effective, for the more vigorously they are waved, the more profusely the *signore* perspire. The organ strikes up, but the chatting accompanied by feverish fanning carries on regardless.

Most of the menfolk have migrated outside to the piazza in front of the church, where they can keep an eye on the service while at the same time smoke and talk in peace. Inside, the hum of voices sinks to a low whisper as the bride enters. Cousin Gianni beams with pride as he escorts his daughter up the aisle, nodding left and right at anyone who catches his eye. Vanessa is beaming too behind her veil, not just nodding, but turning to wave at the guests as she passes. The background conversation continues unabated, competing with the music and the priest's solemn recitations. Nor do the bride and groom feel obliged to observe any outlandish rule of silence, for they happily address each other from time to time during the service. No sooner is the ceremony over than the whole church seems to erupt. The service had been *emozionante*, the floral decorations *elegantissimi*, and the bride *bellissima*, but now it is time to party.

Giusi and Gianni's villa, which faces the tiny beach of San Vito, just half a mile along the coast from Mazara, is famed for its garden. The house, shaded by palms and Mediterranean pines, overlooks a generously sized pool, while surrounding thickets of glossy-leafed shrubs and flora explode in a psychedelic profusion of orange, purple, and pink. In front of the villa, sloping down toward the sea, is that rarity of rarities, a *prato inglese*. Sipping champagne and admiring the verdant carpet beneath our feet, I notice that some of the female guests, unaccustomed to treading on soft surfaces, are having trouble with sinking stiletto heels. Meanwhile we have the task of greeting the whole Vaccara clan and the crème de la crème of Mazara society assembled around us. Eventually we are assigned a table shared with Silvio and Nanette. Not that any of us spend much time actually at the table. There is so much networking to do. In between forays to the buffet table, we sit hosting our space as wave after wave of Family stop by.

The first of the *cugini* to accost us is Agostino, who, like Marcello, spent his early years with the rest of the brood of cousins in the grand Vaccara villa in Mazara. According to one of the Vaccara Aunts, at the tender age of three these two were partners in infantile crime. Substituting salt for sugar behind the chef's back and peeing into saucepans on the roof of the *villa grande* in order to empty the contents onto the Aunts and Uncles on the terrace below are just examples of the exploits accredited to them. Not that anyone here tonight is going to bring up this shady episode in the cousins' distant past. The news that we are to renovate Santa Maria has traveled like lightning. Fiorella, Agostino's sister, is exuberant.

"Marcellino, Caro*line*. So you're making a home at Santa Maria. Are you coming to live in Sicily?"

"*Magari*—if only!" Theirs is indeed wishful thinking.

Like Fiorella and Agostino, Luigi Emilio has come all the way from Palermo. His wife, Enza, arrives quivering with excitement, her eyes brimming with tears of joy.

"Marcellino, Caro*line*. Santa Maria reborn! When is the party? I am invited, aren't I?"

"What party?"

"The housewarming party for Santa Maria!"

"Whoa, Enza! We haven't even started the building work yet."

"Yes, but don't forget to invite me to the party."

As if we would forget to invite Enza. Franzina, Luigi Emilio's elder sister, is more sanguine. With traditional Sicilian pessimism, she reminds us of a local proverb:

Fabbricare casa è un dolce modo d'impoverire—doing up a house is an enjoyable way to lose your money." We know that, we assure her. Everything will be done *in economia*. And then Lallo, the younger brother, gives us his blessing.

"Finally, you've made up your minds."

An occasion such as this brings back memories of the initial impact the Sicilian Family had on me. Meals were the main focus of the day in those early visits to Sicily, as they still are, and most of the day was spent on preparing and consuming them. What was left of the waking hours was devoted to an activity no less important: Getting Close. My first visit to Sicily, when I was on the receiving end of boundless, exquisite Sicilian hospitality, bore little relationship to my second, when as a sworn-in member, I had to assume the duties and responsibilities of one of the Family.

Marcello had already provided me with a family tree of sorts, but it was so vast and its branches so extended that it had taken on the proportions of a jungle, impossible to commit to memory. The task in hand was certainly not helped by the fact that both his grandfathers, Vaccara and Manzo, had been called Luigi. The rule is, as Marcello explained, that the eldest son is named after the paternal grandfather, the eldest daughter after the paternal grandmother. This was why he had so many cousins called Luigi, six to be exact. There didn't seem to be much flexibility to this rule. Only Pilli had managed to escape the Luigi nomenclature, partly because by the time he was born, there was already a surfeit of Luigis in the family, and his patriarchal grandfather, Luigi Manzo, the notary, being already dead, was in no position to protest any heretical change-of-name proposals.

"So, does that mean when we have children, we'll have to call the first boy Goffredo and the first girl Maria?"

"*È logico.*" I didn't feel too happy about that, but since there was no little baby Manzo on the way at the time, I decided I'd deal with the problem when it came up. The pressing issue just then was to come to grips with the family tree.

"With six cousins all called Luigi, don't you get them mixed up?"

"No, it's easy. There's Luigi Emilio, Luigi Umberto, Luigi Andrea, Luigi this, Luigi that . . ."

But as I soon learned, these were names used only to face. When referring to the various Luigis behind their backs, the family reverted to what Sicilians call *l'ingiuria*—the nickname. But the literal translation of *ingiuria* would be "insult," and indeed none of the nicknames are complimentary. Luigi Umberto is known as "the Baron," a reference to his mother's aristocratic lineage. (When I first met him it took but one look to see why the nickname had stuck. He was short, bald, and stocky, with a wide gappy grin and apparently lashless blue eyes exuding an obviously boundless good nature. The joke was in the contrast between the grand title and the country bumpkin appearance.) One of the Luigis on the Manzo side of the family was dubbed "*Pilastro*"—the pillar—because of his stiff gait, yet another, Luigi, "*Coniglio,*" because of an unfortunate investment he had made in a rabbit-breeding farm. And just as each eldest male Vaccara was called Luigi after their grandfather, the eldest girls, Franzine, Franca, and Francesca were named after their paternal grandmother.

"And the ingiuria? Why do none of the women have an ingiuria?" Marcello gave this question some thought before replying.

"I think it's because we respect women too much to give them an ingiuria. The ingiuria is only for men."

IF A FORESTIERA IS TO BE ACCEPTED INTO THE MANZO VACCARA clan, then she must be educated in their ways. Once Marcello had brought me back to Sicily as his fidanzata. I had to be introduced to all available members of the extended Family, from Maria's and Goffredo's numerous siblings, the aunts and uncles—the *Zii* and the *Zie*—through to countless ranks of cousins, their spouses and in-laws, friends of long standing who had been granted honorary status, and anyone who had ever worked for the Family. This was a task that Maria nobly took upon herself. Given the limited time available—just under a week—the undertaking required strategic capacity and organizational skills of a very high order. But my mother-in-law was ready and able to meet the challenge. Once the campaign was under way, Maria, an indefatigable and boundless source of energy, did not let her protégés out of her sight for a second. Time was short, and she was a Mamma with a mission. Marcello must have had some inkling of what we were in for, but being Sicilian, and therefore a sociable animal, he was far better equipped than I was to deal with it. In his breezy way, he somehow forgot to explain to me in advance just how much pressure we were going to be under. As soon as breakfast, lunch, or dinner was out of the way, Maria would summon us.

"Caro*line*, Marcellino! Are you ready? *Andiamo!*" Blighted with a fair amount of English reserve in addition to old-fashioned shyness of my own making, I must have been unpromising material for Maria. Whether or not she was disappointed in my performance, it soon became clear that she was making no concessions to a trait in her new daughter-in-law that she found incomprehensible. Within days I was in muted revolt, begging Marcello to intercede with his mother for some free time for the two of us—an hour or two, just to walk in the country or stroll down the promenade or the beach. To humor me he agreed to arrange a brief respite, but not via the official channel. There was no point in trying to explain my need for privacy to Maria, as she would never have understood it, much less agreed to indulge it. Subterfuge was the only option. After lunch one day, while Maria was still in the kitchen, we sneaked downstairs and out to the car. It's hard to start the engine of a Fiat Cinquecento without making a noise, and sure enough, it took no more than a couple of chugs to alert Maria to the escape attempt. Quick as a flash, she popped her head out of the kitchen window and caught us in flagrante. With a gleeful shout, she hurried down to the courtyard, delighted at the prospect of another outing. Seeing no point in going to an empty beach, quite frankly, she redirected us to the villa of a friend of hers, and suddenly we were on the visiting trail again.

Slow to recognize the compulsory nature of Maria's schedule, at the beginning I had even had ideas about Marcello and myself taking a day off for a romantic excursion on our own, such as a picnic lunch to the beach perhaps or a sightseeing trip to Erice or Palermo. But I soon discovered that this was not to be. Individual initiative was disruptive. If we went anywhere on our own, then Maria would have to come too, she explained, and that would upset everyone else. In any case, it was not how things were done. Nobody stayed on the beach after one o'clock. Why would we want to go without a proper lunch? Who was making us do that? And why on earth would we want to spend hours in a hot car when we could stay here in comfort with her and the rest of the Family? If we really wanted to go anywhere, we should wait until after lunch. Since lunch was never over before four, and compulsory Getting Close was already under way by five, this was hardly feasible. The novelty of joining the Family began to wear off as I took on the obligations that went with membership. It was impossible to go anywhere or do anything without consulting my mother-in-law. Consultation led to objection, and inevitably my timid proposals were doomed to founder, leaving everyone else free to get on with lives revolving around Food and Family, and me a truculent captive in a run-down castle. I was beginning to see the dark side of my love affair with Sicily. When I protested in private to Marcello, he reminded me again with a sigh, "Didn't I tell you my mother was prepotente?"

Unfortunately for Maria and fortunately for me, Goffredo's family, the Manzo clan, was based in Trapani, an hour's drive away—too far for regular visiting. That would have to take place at a later date. Meanwhile, Maria concentrated her resources on a frontal assault on her own family, the Vaccaras. By the midseventies, when I arrived on the scene, my mother-in-law's generation had already been depleted. Her eldest brother, Nino, the *eccentrico* of the family, had died of cirrhosis of the liver while still in his forties. Stefano, l'onorevole, father of Franzina, Luigi Emilio, and Lallo, had had a glamorous career as an entrepreneur and member of the Sicilian Parliament, only to spend his last years in reduced circumstances. That left Zio Giovanni and Zio Franz. I always preferred the visit to Zio Franz; his household was more informal, and

I loved his Cheshire cat grin, inherited by all three of his children. However, in the Family hierarchy, Zio Giovanni, the elder of the two brothers, took precedence in Getting Close.

Giovanni was the only Vaccara to have held on to his father's inheritance. He and his wife, Zia Checchina, still lived in the grand Vaccara palazzo in Via Maccagnone. It had a daunting atmosphere clearly designed to impress: heavy velvet curtains in ugly burnt umber, highly polished marble floors in tombstone gray, oppressively dark furniture, Venetian mirrors, gloomy portraits, and outsize paintings of tortured saints in ornate gilt frames. When we arrived on an official visit, Zio Giovanni was at work at the Vaccara wine factory and Zia Checchina was playing poker, as she did every afternoon. While we waited for the game to finish, Maria told me a story about Checchina's sister. When in a benevolent mood she would offer her maids *cioccolatini*, but instead of offering them on a platter, like a feudal baroness, she would throw them on the kitchen floor and watch the serfs scramble like dogs. And this took place in the twentieth century. Maria was into another story about a maid who had been badly treated and systematically spat into her sister-in-law's coffee every morning for years, when our hostess finally came in, pleased to have won the game of poker. For all the blue blood in her veins, Zia Checchina was no beauty, but despite the pop eyes and stocky build, she had a commanding presence. Even Maria seemed slightly intimidated. Our hostess invited, or rather ordered me to come and sit next to her.

"Carolina, come here and make me laugh." To the Zia Checchina I was a comic figure, guaranteed to provide a laugh or two. I protested that I didn't know any jokes I could tell in Italian.

"Oh no. That's not what I meant. It's just the way you speak that's so funny." And I had been so pleased with my progress in mastering the language.

A uniformed maid, young and jolly looking, definitely not the spitting type appeared with a decanter of Marsala and teeth-cracking *mustazzoli*—must biscuits made with grape concentrate—the obligatory accompaniment. Before I ever came to Sicily, I had been warned by friends in Milan.

"Watch out for those Sicilians. They'll be giving you Marsala for breakfast."

If the decanter never actually appeared at the breakfast table, Marsala was certainly on offer by eleven in the morning, which was fine by me. After all, if one did have to devote such a large proportion of the day to Getting Close, Marsala was a far more interesting alternative to morning coffee or afternoon tea. I had to be careful not to show my appreciation too vigorously though. If Maria was any example, a signora was supposed to renounce this elixir of a drink, while the male members of the family, like Marcello, were at liberty to down it with the enthusiasm it deserved.

Zia Checchina announced that she had a wedding present for us, actually for me. Having heard Maria's anecdotes, it did occur to me that the Zia might fling her offering on the floor. Fortunately she did no such thing. Since we were about to migrate to Latin America like Paddington Bear with only a suitcase for luggage, she had had the forethought to make the gift portable. Expressing a heartfelt *simpatia* for the inglesina, she placed a small gift-wrapped box firmly in my hands. It turned out to contain an 18-karat gold watch. I was delighted, but my effusive thanks in my best Italian only had the effect of reducing Marcello's aunt to giggles. As soon as we were out of earshot, Maria sniffed at what she considered to be a downmarket brand—with all that money! In any case, my enjoyment of this expensive trophy was short-lived, for only months later a thief on a Milanese tram relieved me of the most valuable present I had ever had in my life.

ANYONE WHO APPRECIATES GOOD FOOD CANNOT ATTEND A Sicilian wedding without paying each dish on offer the tribute it deserves. Vanessa's wedding presents Marcello and his brothers with an apparently insuperable challenge. The spread is so vast and so varied that not one of them manages to sample *every* single dish.

"*Troppe cose*—too many things to eat! *Troppo buono*—too good for words! *Come fa un cristiano ad assaggiare tutto*—how on earth can a Christian (Sicilian for the man in the street) be expected to savor everything?" By the time we have made our way through three generous courses of seafood antipasto, pasta with swordfish, and a wide assortment of fish, there would seem to be no room left for *dolci*. Yet these are Sicilian dolci, and it goes without saying that no one can resist at least a taste. Clem's eyes light up at the sight of cannoli, horn-shaped shells of crisp pastry filled with an exquisite blend of fresh ricotta sweetened with sugar and candied orange peel. There are tiny almond pastries, watermelon jellies dotted with chocolate flakes, and great logs of elaborately decorated ice cream, but to my mind all are outshone by a giant snowy *cassata,* that triumphal amalgam of marzipan, sponge cake, and sweetened ricotta all swathed in white icing. This one is beautifully decorated with strips of green marzipan and candied fruit, the pears and cherries cut to resemble the petals of a flower and garnished with translucent bands of pale green candied squash.

As I gaze at the spread of dolci set out before us, I can't help saying a silent word of thanks to the Arab conquerors, who brought us these delights—even ice cream, which they called *sciarbat*—sherbet—and made with snow collected from the slopes of Mount Etna. Again, it was the Arabs who brought the sugarcane to Sicily along with citrus fruit, cinnamon, pistachios, and the recipe for marzipan, and it is to the Arabs that Sicilians owe their sweet tooth and hence a reputation for exotic and wonderful dolci.

While the Manzo men rail at the fact that their digestive systems simply cannot cope with the abundance, my sisters-in-law flit back and forth between tables, ceaselessly networking as if they are about to be transported to the Antipodes and will never set eyes on the rest of the Family again. Clem, like her father, ever lured down the path of gastronomic temptation, and like her mother, ever anxious that the resulting pleasure will not distort her figure, claims to be so full she's going to

burst. She persuades me to take a walk around the garden. The whole
Vaccara clan and some of the Manzos too are gathered at these tables set
around a pool shimmering with light. By now Vanessa's friends have
abandoned smart clothes and the buffet table for swimming gear and
the pool. Clem is amazed at the informality of it all.

"Is dinner over then? No toasts? No speeches?"

"Obviously not. But I think they've got their priorities right, don't
you? Cutting out the formalities means there's more time for what
really counts: eating and socializing."

We bump into a dripping wet Dario, wandering around looking for
his bride.

"Have either of you seen a beautiful girl called Vanessa?"

The moon is full, and the night thick with the hum of conversation. Snatches of talk drift our way through the palm trees: *"Che bella è
Mazara!"* comes from a group of Vanessa's cousins. I recognize Raffaella, a smiley, effervescent mother of two, whom I once bounced on
my knee as a six-year-old.

"I love Mazara," she announces to the world. "I will never leave it,
ever, ever, ever!"

It is hard to imagine English twenty-somethings singing the praises
of their home town. Clem, who overhears Raffaella's exclamation,
raises an eyebrow in disbelief. She is deeply attached to Sicily, yet having been brought up in exotic climes and used to traveling all over the
world, she finds the idea of being confined to a small provincial town
for the rest of her natural life dismaying. At times Marcello and I feel
guilty for not having provided the roots that children apparently need,
but at other times it seems that the Flowers shrug off the very roots we
have failed to provide, fitting more easily into the role we have imposed
upon them of citizens of the world. Is it their English blood that sepa-
rates our daughters from a Sicily in which love of one's home town and
all things Sicilian is taken for granted? For Maria's generation, the only
reason for leaving home was economic duress. It was only when family
fortunes plummeted that Vaccaras had to leave Sicily and go to the

continente—the mainland. Marcello and Pilli were the first to do so, and although economic reasons have made it impossible for either to live here, both have returned to Mazara every year without fail, and their children's link to Mazara, although more tenuous than that of their Vaccara cousins, is still there.

Almost a quarter of a century has gone by since I was escorted around Mazara on a compulsory Getting Close campaign, and the Vaccara uncles and aunts have now all died, survived only by Maria herself. My mother-in-law would have reveled in a Family wedding such as this, but now in an advanced stage of Alzheimer's, she is unable to attend. Looking around we realize that with the exception of Maria and one of the Luigi's, every single member of three generations of Vaccaras is present. Clem wonders at the organizational skills required for such a feat.

"How on earth did Giusi manage to get them all together?"

"Easy, really. Most of them live here anyway, and those who don't come to Sicily every summer, just like us. It's part of what they call *sicilitudine*."

CAROLINE, 1976

SIX

Carpe Diem

MY FIRST VISIT TO SICILY INITIATED ME INTO THE CULT OF *mangiare*, but it was only after marriage to Marcello that I learned of its pivotal importance to the Sicilian soul. Food and thoughts of food may dominate one's waking hours. This I could understand, but in Marcello's case the subject also haunted his dreams. His grandfather and uncles had employed a series of cooks of the genre known as *monzù*—monsieur. Originally from France, these cooks arrived in Sicily with the establishment of the Bourbon regime in the eighteenth century and soon became a must-have for aristocratic families. By the twentieth century the average monzù, like those in the Vaccara employ, had never been farther than Naples. Marcello was a frugal eater, but if he had had a magic lantern, his first wish for the genie would have been for one of these live-in professional cooks, ready to conjure up exquisite dishes at his command. From time to time he would stumble bleary-eyed to the breakfast table, reporting, "I dreamed of Don Pedro's consommé. It was *squisito!*" Why, I wondered, was he so obsessed with these doyens of a bygone age? And were his dreams a reflection on the harsh reality of English cooking he was forced to endure? Ah well, I comforted myself, if they were, then the punishment had been self-imposed. He had only himself to blame.

Now of course, the monzùs are very much the stuff of dreams, well beyond the means of the average middle-class family. Although Maria and Goffredo were brought up to expect other people to do all the manual labor, including the preparation of food, both liked to apply a hands-on approach in the kitchen. Goffredo was an acknowledged gourmet, an inspired cook, but *inspiration* was the key word; he ventured into the kitchen only when inspiration took hold. I heard many stories of his accomplishments but was told I would have to wait months, perhaps years, before he decided to treat the family to his famed veal galantine or one of his exquisite fruit sorbets. On the other hand, Maria excelled in the perfect execution of traditional Sicilian fare. No doubt her mother, Nonna Francesca, who had always stayed close to her contadino roots, had influenced her daughter's passion for simple food, especially *salsa*—tomato sauce—the cornerstone of Sicilian cuisine.

While Maria and Goffredo were both accomplished cooks, there is no doubt that in this field the reputation of the parents was totally eclipsed by that of their eldest son. It was not until Marcello and I were already married that I first met Pilli, the hero of many family tales. Our wedding had been a hasty civil ceremony in England, and there was no time for a proper wedding, let alone a honeymoon. Or was there? Italian and Brazilian bureaucracy combined to bungle our visa applications for Brazil. The issue was complicated, and it looked as though we might have to wait in Italy—with nothing to do—for months on end. No point in hanging around the office in Milan, Marcello decided. What would be the point of that? Much better to go back to Santa Maria. So my first meeting with Pilli took place at Rome airport, from where we were all three bound for Sicily.

Having seen the photos, I had already discerned Pilli's good looks in the style of Marcello Mastroianni. But Pilli was shorter than his brothers, with a distinct potbelly discreetly concealed by a safari jacket. Despite the flamboyant dark hair curling over his collar, the Manzo mustache, and goatee beard, he still managed to be the spitting image of Maria. The two brothers embraced and exchanged kisses on both cheeks, then talked at each other non-stop throughout the flight to Palermo. As I observed my brother-in-law, I tried to match the man with the myth. I had been told he was an extrovert, top of the class at school and university, the life and soul of the party, a lady-killer, avowed Communist, gourmet, and gifted cook. With a personality as forceful as his mother's, he had always overshadowed his more temperate younger brothers. The strong likeness to Maria was not just in the facial features but in his effervescence. With his knack of taking over and dominating the talk, he was like a bubbling pan that wouldn't take a lid. His *allegria* was infectious, of the kind that makes you laugh even when you don't get the joke. Like his mother he was full of projects. He was going to give up the *sacrificio* of building bridges and dams in Africa and start up an import-export business instead. He had plans for Santa Maria too, to convert a wing of the villa and settle the family there. He flirted with the idea of an avocado plantation, a restaurant, and even a hotel. On arrival I met Maria Antonietta, his wife. Despite the mouthful of the old-fashioned name, she was snappily dressed and strikingly attractive. Petite and vivacious, with wavy blonde locks and a husky voice, she had the glamour of Monica Vitti, a perfect match to Pilli's (slightly paunchy) version of Marcello Mastroianni.

While Maria commanded the troops in the regular Getting Close maneuvers, she stepped aside when it came to the most serious matter of the day: *mangiare.* It went without saying that the three generations assembled at Santa Maria would come together regularly for lunch and dinner. Under Pilli's direction, however, the planning and preparation of meals was highly organized and more demanding than it had been on previous occasions. It started with a breakfast conference.

Scenario: (early morning in the upstairs kitchen) Maria, the materfamilias, and Antonietta, the maid, having a Moka espresso together.

Meeting objective: to decide what to have for lunch, and who is making what for dinner. The women wait for the men to join the meeting before any executive decisions can be made. Pilli, chairman and creative director, arrives first on the scene, followed by Goffredo and Silvio. Marcello puts in an occasional but distracted appearance. (He is a late developer in the school of *mangiare* and has not yet learned to give the subject the concentration it requires.) After the cross-table talks in the kitchen, the executive committee moves to the dining room to draw up a strategy. Pilli takes the chair while Maria, seated on his right, doubles as coordinator and PR executive. With Goffredo and Silvio acting as consultants, the committee thrashes out the options until they come up with a final list of recipes. There is no consulting of cookbooks (if there were such things in the house, I never saw them) or taking of notes. There is no need to consult the written word when the recipes are as well known to everyone as family birthdays or multiplication tables. When they finally come up with an action plan of which ingredients to purchase, where and by whom, no one bothers to draw up a list. Each member of the committee takes a mental note of the assignment allocated to him. As for quantity, somehow they just know how much to buy.

Initially I had been too overwhelmed by the place and the people to give careful attention to what I was eating. Once I was officially one of the Family, I could relax in my efforts to make a good impression and pay the food the respect it deserved. Sicilian cuisine is celebrated for its diversity and sophistication, but the main object of Sicilian culinary passion, at least in this family, was la salsa—tomato sauce. Life without it would be unthinkable, and Heaven only knows how Sicilians managed to survive before Columbus came back from the New World with the tomato. Poor Marcello had stoically done without the real thing for months, ever since he married me, in fact. Unlike his brothers he had not learned to cook, and in his bachelor days he had always dined in restaurants. Consequently the experience of eating at home where the cook was unskilled and English had come as something of a shock. Not that I hadn't done my best to please my new and exacting husband. With the help of Elizabeth David's recipes, I had tried my hand at any pasta dishes that looked simple—alla matriciana or with a mascarpone sauce—taking care to serve the pasta al dente. These little triumphs earned praise from Marcello at the beginning, but even so, I was aware that even if not intended to damn, the praise was on the faint side. What my Sicilian husband really pined for was a plate of spaghetti with tomato sauce, the way his mother and Antonietta made it. But when I attempted to make the beloved salsa, somehow the results did not come up to standard. The point was brought painfully home one evening in a restaurant in Milan. Marcello ordered *spaghetti al pomodoro* for the first course and then again for the second.

"Don't be ridiculous," I hissed. "Order something else. You can get pasta with tomato sauce at home!" I meant home, *our* home, the one he shared with me (not Sicily, of course).

"Yes," with a sly glance at the waiter, "but at home it's not the same."

My first extended summer visit to Santa Maria was a welcome opportunity to come to grips with the salsa. Armed with pen and paper one morning I asked Antonietta to teach me how to make it. The first cooking lesson of my life took place in the heat of the upstairs kitchen, watching Antonietta's hunched figure, clad in black, shuffling between the blackened walls in her apron and slippers. While she gabbled away in thick dialect, Maria flitted in and out, occasionally acting as interpreter. Unfortunately, during the translation process the mistress would contradict the maid and proffer her own version of the right way to make salsa, much to Antonietta's annoyance. In the end they both lost interest in the cooking lesson and started arguing over the recipe. As for quantities—how many tomatoes? how much basil? how much garlic?—both of them took the same cavalier attitude. The answer was always "*assai*—a lot," a word frequently used by those used to cooking for large numbers and generous appetites. Despite the language difficulty, the heat, and the arguing, I eventually gleaned the basic principle of a good salsa, which is to simmer the tomatoes first and add the flavoring later. As for the tomatoes, mistress and maid took a united stand on this point: they had to be the *pizzutelli*, the tiny thick-skinned variety that matures in late summer. The plum tomatoes that they use up north were inferior. Salsa tomatoes should be firm with a minimum of water content, just like Sicilian tomatoes. Tinned tomatoes most definitely would *not* do.

What is it that makes the Sicilian tomato so full of flavor? The soil is extremely fertile and the climate distinctly tomato friendly. The plant gets all the water it needs in short sharp bursts in the mildest of winters, then nothing but sunshine for six months while the fruit ripens on the vine until it is harvested in August. All this makes for the water-free, full flavored flesh typical of the Sicilian pizzutello.

Almost as indispensable to the Sicilian soul is *la melanzana*—the aubergine, or eggplant, preferably the Tunisian aubergine, huge, round, and lilac colored, not the elongated deep purple variety usually available in British supermarkets. The aubergine seems to appear at every meal. Puréed, stuffed, or fried in huge great disks to go on top of pasta, it is a natural partner for the tomato. These were the main ingredients of one of Maria's greatest culinary triumphs, *caponata,* a velvety dark and viscous vegetarian concoction that she prepared once a year and kept in jars to be brought out on special occasions. Caponata is not immediately pleasing to the eye, which makes the delicious taste so surprising. Maria gave me her recipe, but I never managed to produce a caponata to rival hers. Her instructions involved hours of chopping and frying aubergines before combining them with onions and celery (more frying) and tomato sauce. Vinegar and sugar were added to produce a taste at the same time both sweet and savory. All this was enriched by the addition of olives, capers, and toasted almonds. The result was so extraordinary; it simply had to be tasted to be believed. Aubergines and tomato were also predominant in Marcello's favorite dish—*melanzane alla parmigiana*—fried aubergines with lashings of salsa and fresh basil and a light sprinkling of grated Parmesan. Once I had mastered the salsa, I would be qualified to try my hand at this too. Most important, I was told, is the temperature at which it is served—cold, not just luke-warm, but cold. Only this way do you savor the contrast between the pungency of the ripe tomatoes and the delicate, smoky tenderness of the fried aubergine. Though relatively simple to prepare, it was the subject of one of many controversies about food. In this case the dispute was about the method of frying, whether the aubergines should be soft and dripping with oil after frying, or crisp and dry. The soft and sodden school won the day when Pilli came down in their favor. Nothing worse than a *Parmigiana* ruined by dried up aubergines, he insisted.

The ultimate treat was Antonietta's memorable *pasta al forno*, the same dish she had made in my honor on my first visit. This queen of pasta dishes took up to a whole day to prepare, so it was usually served in the evening and did not often appear in the height of summer, when the temperature was often over 100 degrees Fahrenheit, higher still over a hot stove. Although it is basically oven-baked pasta, it is only a very distant relation of the *lasagna* of the Emilian region in the north. The difference is that Sicilians dispense with the béchamel, which makes it less subtle but so substantial that it does for both *primo* and *secondo*—pasta and main course. The vital ingredient here is *ragù*, a concentrated sauce made from tomatoes, sausage meat, and veal. Absolutely essential is the *stratto*, a thick sticky paste extracted from sun-dried tomatoes. As for the pasta itself, it can be any short pasta, but the cooks of Santa Maria prefer to make it with *anelletti*, the little curly rings that look like pigs' tails and become deliciously crispy when baked. The pasta—only half-cooked to allow for baking time—sauce, and fried aubergines are placed in layers to fill a mold, and, as if this were not rich enough, the dish is spiced with cloves, pine nuts, sultanas, cheese, and basil. Its consistency is firmer than that of lasagna, so the ragù must be dryer than a standard *bolognese* sauce. After baking, the pasta must be turned out, and there is always that moment of tension when it slithers out of the upturned mould onto the ceramic tray. Will it hold firm, or will it split its golden crust and slump into a runny mess? With expert hands like Antonietta's, it never slumped but always stood gloriously firm in all its gargantuan splendor—a steaming mountain plateau of solid comfort food, fragrant with spices, a triumph of Sicilian abundance.

Maria introduced me to many an exotic pasta dish that first summer, but one of the most surprising was *pasta con le acciughe*—pasta with anchovies. Here, in one of the most humble but delicious dishes imaginable, the spaghetti comes dressed in a simple sauce of salted anchovies melted in a generous quantity of olive oil with a little sautéed garlic and chopped parsley. The dish is topped with *pangrattato*—homemade breadcrumbs—the poor man's cheese, in this case fried to a crisp golden brown and ladled generously over the plate. This is a typical example of how Sicilians take the poorest ingredients—anchovies and a hunk of stale bread—to come up with a concoction that is nothing short of sublime.

But the one ingredient I have not mentioned was the most important of all—*il pane*. Whereas I had always thought of bread as an accessory, not a staple, essential in sliced form for breakfast toast, inevitable in sandwiches, but otherwise dispensable, for my new family, as I soon discovered, it was literally the staff of life. Never had I come across people so attached to their bread. The days when it was made in the wood oven in the downstairs kitchen were long past. Now it was bought daily in two trips to town, one before lunch and one in the late afternoon, which guaranteed that we would have freshly baked bread at both meals. It was never put in the fridge, and the leftovers never went into a freezer (there wasn't one) but were kept to make the breadcrumbs that are such an important ingredient in so many dishes.

All Italians love bread, but true to the Goethean principle, Sicilians are simply passionate about it. "A day without bread is like a day without sunshine" according to a Sicilian proverb, both ingredients being an essential to Sicilian life. There is an Italian word for any food that is eaten with bread—such as ham, cheese, meat, and vegetables—*companatico*. As Maria often reminded us, one can eat bread without companatico, but companatico without bread is unthinkable. It was an absolute must at every course at every meal, and Maria always made sure that we had plenty of it. We snacked on it with *passuluni*—wrinkly black olives—or salted tuna and pecorino cheese, while we waited for the pasta. We used bread instead of a knife to eat fish, and we even ate it with fruit. What is all the fuss about with Sicilian bread? Sociologists argue that it is a symbol of survival. In a society where famine was an omnipresent threat, if there was bread on the table, hunger was at bay.

They may be right, but there is more to it than that. Famine was known all over Italy during the Middle Ages but the differences in types of bread are as great as the cultural differences between the regions. Compare the Milanese *michetta*—the bread roll—which resembles nothing so much as a chunk of polystyrene, or the tasteless bread of Umbria, still made without salt, which papal taxes had once made unaffordable. Sicilian bread is more substantial than any other in Italy—the *ciabatta* from Puglia coming a close second—and delicious to the point of addiction. Like all Sicilian food, the bread has a strong flavor, once tasted, never forgotten, particularly the deep brown crust, topped with sesame seeds, a delicious legacy of the Arab occupation. But it is the flour itself which puts Sicilian bread above all rivals. The semolina flour is milled from the durum wheat that grows only in southern Italy and Sicily, the same silky flour that makes Italian pasta unique. In ancient times, wheat was Sicily's main crop, and it was the fertility of the soil that made the island so attractive to invaders in the first place. By the third century BC, the Romans were casting greedy eyes on the fertile plains of the neighboring island, with its famous crop of *triticum durum* and a harvest almost three times higher than on the mainland. With the conquest of Sicily, Rome had access to that harvest, and before long she was dependent on Sicilian grain to feed both the army and the fast growing populace of the capital. Cato the Censor coined the phrase *nutricem plebis Romanae*. Sicily was the nurse at whose breast the Roman people were fed. The Roman army marched on Sicilian grain, and the bread of the "bread and circuses" that filled the stomachs of the Roman plebs came from Sicily.

Lunch at Santa Maria is usually served between two thirty and three in the upstairs dining room. By the time Marcello and I were married, Antonietta no longer served at the table, although she still worked behind the scenes. Her waiting duties had been taken over by a newcomer. Silvio had been in Togo on a government contract for a brief spell and had come back with his Togolese houseboy, who went by the name of Antoine. Now installed in the ex-servants' quarters, in theory he should have been working as a team with his namesake, but in fact Antonietta treated him like dirt. Communication was problematic, since she knew only Sicilian dialect and Antoine spoke only French and Togolese. The lack of communication was probably in his interest however, as it sheltered him from the barrage of insults directed at him by his colleague, whose ferocity more than compensated for her diminutive stature. He got the general message, and despite his proven competence as a cook, he was wise enough to keep out of the kitchen until it was time to serve lunch. For this most important duty Maria had dressed him up in a livery of black trousers, red jacket with gold embroidery, and white gloves. Harking back to the age of the monzù and thrilled to resurrect bygone grandeur, she had trained the new butler to serve meals in the manner to which she had once been accustomed. She would reel off her instructions, showing off her fluent French while Antoine murmured back a respectful "*Oui, Madame*" as he glided around the table, balancing a huge platter of pasta on his arm. Patiently he navigated the sea of gesticulating adults, not to mention Pilli's children, who regarded the dining table as an activity center, where the last thing they were expected to do was sit still.

Silvio had been worried that Antoine would be socially isolated in a provincial Sicilian town. Marooned in the country and unable to speak Italian, he would find it hard to make friends. But within days Antoine had discovered the French-speaking community of Tunisian immigrants, and once he had acquired a smattering of Italian and a secondhand Vespa, his social life was launched. Since his duties at Santa Maria were over by sunset, every evening he revved through the gates heading for Mazara, dressed to kill in slinky black T-shirt, tight black jeans, and a large gold medallion dangling on a chain around his shiny oiled neck.

Like Getting Close, shopping duties seemed to take up a disproportionate slice of the day, but whereas in the past food purchasing had been somewhat haphazard, under Pilli's regime it was strictly regulated. Maria would pick up bread, salad, and fruit, but the serious shopping— trips to the butcher or the fish market—were for the men. Whereas the whole family would contribute to the evening meal, the preparation of lunch was usually left to Antonietta and Maria, and the male members of the family would visit the kitchen only to taste and offer advice, open a bottle of wine, or nibble at cheese and olives. Lunch was usually an intimate affair, with just Maria, Goffredo, and Silvio, Pilli and Maria Antonietta and their children, and Marcello and myself.

"*Mangiachetifabene! Mangia! Mangia!*" This exhortation to "Eat, eat, eat—it does you good," was a constant refrain. It was usually directed at the children but also at me whenever I showed any signs of slacking at the dining table. Although lunch was a three-course meal, consisting of pasta, main course and fruit, supplemented by various side dishes, not to mention the addictive Sicilian bread with its delicious sesame seed crust, everybody seemed to take the superabundance in their stride and partake of everything on offer. As a new addition to the family, I was still treated like a guest, which meant not only could I refuse nothing that I was offered, but for some inexplicable reason, I was expected to accept extra large portions, seconds, and sometimes thirds. Failure to do so aroused concern.

"What's the matter Caro*line?* Why aren't you eating? Are you ill?"

Maria, as the materfamilias, was the most insistent, but at times Goffredo, Maria Antonietta, and even Antonietta, the maid would join in the refrain. Of course I wasn't ill, and if they had been watching me closely, they would have seen that I was doing my best to do justice to the magnificent portions ladled onto my plate. How could anyone possibly claim I was not eating? And how did they manage to stay so slim—well, the women anyway? Of the team of vigilantes my mother-in-law was the most difficult to deal with. Had I not been convinced that she was acting out of Sicilian hospitality and genuine concern, I might have suspected that she was maliciously trying to fatten me up. To add to my embarrassment, she developed a disconcerting habit of addressing me in French, which I could hardly speak. Could she really have been confusing me with the Togolese butler? Granted, we were both foreigners, but even so . . .

The Manzos were storytellers, every one of them, and despite my uneasy relationship with the force-feeding system, I looked forward to mealtimes for the anecdotes about family members. Goffredo was the *raconteur par excellence*. He had a habit of recalling episodes that I found somewhat disconcerting.

"Did you know there has been a murder in the Manzo family?"

"You mean the Mafia murder in Trapani?" But this is a well-known fact. Why is he bringing that up now?

"No, long before that. It was in the 1920s. Sara, widow of the notary, Francesco Manzo, was murdered in her own house by a maid and her lover. They attacked her in her bed, but she was quite an old warhorse. She must have put up a fight. The incredible part is that her grandson Ciccio (short for Francesco, Goffredo's eldest brother) was in the house in the room below while it was happening. He was busy practising a speech for an appearance in court the next day and was so absorbed in it that either he didn't hear the noise from upstairs, or, as his grandmother was always arguing with the servants, he just ignored it and went to bed. The next day he found she had been stabbed to death, and he himself was the prime suspect."

"And the murderers?"

"They got them in the end, to Ciccio's great relief. The lover turned out to be a deserter from the army with a criminal record."

Poor Sara. She was the only woman in the family to be afflicted by the Manzo temper.

Ah the Manzo Temper! This had always been considered a male prerogative. When it cropped up in a woman, she came to a bad end. Goffredo was one of the most renowned exponents, but he did not just go in for the odd fit of uncontrollable rage. When it came to displays of temper he had flair and imagination. Always a gentleman, even when in a fury, his temper would immediately give way to violence of an unorthodox variety. It flared up on one occasion when a Limoges dinner service was delivered to the door. Maria had ordered it without his knowledge at a time when they could ill afford such luxuries. White-hot with rage and without uttering a word, he ripped open the packing case, wrenched out the precious porcelain plates, unwrapped each one, and hurled them one after another against the wall as Maria stood by helplessly. There was no stopping him, and indeed, she knew better than to challenge him once the Manzo Temper had seized. In the end, to get the whole thing over with as soon as possible, she joined in the exorcist operation and calmly handed him piece after piece until the whole set lay smashed to smithereens at their feet. This was in the fifties, long before normal people went in for public displays of anger or road rage. Indeed my father-in-law would never have descended to bad language or fisticuffs. His violence always took a more eccentric and aristocratic form. On another occasion, after negotiating a wine sale with a client of dubious extraction, he realized that he had been tricked, but it was too late to do anything about it; the contract had already been signed. So what did Goffredo do? Instead of resorting to verbal abuse, he quietly removed his hat, dropped it to the floor, and, as though the hat were the waxen image of the man in front of him, stamped on it in silent fury until it was reduced to shreds. No doubt the trickster got the message.

If the Manzo side of the Family is afflicted by Temper, by common consent the Vaccaras are of less irascible stock. I was curious about Maria's father, Luigi, the patriacrch. A brief glimpse of him on a film taken in the family home in 1942 certainly does not give an impression of geniality. The old film shows the whole Vaccara clan assembled in the garden of the grand house in Mazara: Luigi and his wife Francesca, their five married children and the grandchildren, a host of little Luigis and Francescas. Marcello was a baby in Maria's arms, Pilli a toddler and Silvio was not yet born. The scenes would be of little interest to anyone but the family concerned—apart from a riotous Billy Bunterish scene where, in a vain attempt to retrieve a length of stolen sausage, the cook chases a fat and lumbering Luigi Umberto (the future Baron) round and round the fountain. The handsome, youthful Goffredo is the most active of all the parents, the only one who actually appears to be paying any attention to the children. There's no sign whatsoever of the hat-crushing,

plate-hurling, middle-aged Manzo Temper. Don Luigi wears a permanent scowl, not in the least amused by his grandchildren's antics. At one
point he even waves his stick at them. If he was so dour with children,
what kind of taskmaster could he have been? Certainly he was a tough
negotiator and perhaps not as universally popular as Maria claimed. A
local rhyme about him goes,

"*Mangio spaghetti senza olio*
Ma denaro di Vaccara non lo voglio
I'll eat spaghetti without oil,
But I won't take Vaccara money."

On the other hand, Maria, who doted on her father, never fails to
remind us that Luigi Vaccara treated his employees exceptionally well.
In an era when Sicilian farmhands were the poorest in Italy, he offered
better wages and living conditions than other local employers and was
famous for his consideration and generosity. He is certainly remembered with affection by hundreds of ex-employees and their families.
Indeed, I have heard too many ex-retainers reminiscing with tears in
their eyes over the good old days in the employ of Don Luigi to doubt
this verdict.

Daytime, with the heavy midday meal and the enforced rigors of
Getting Close, could be something of a trial for a new recruit to the
Family, but when the sun went down, the mood relaxed and all trials
were forgotten. By evening the scene had shifted from the upstairs dining room to the *cortile*, the beautiful leafy courtyard at the center of the
villa, swathed in those days in bougainvillea, jasmine, and canopies of
drooping ivy. The barbecue was used every night, and it did seem like a
fair and sensible way of catering. Once the women had served the
pasta, the men took over, as they do at barbecues the world over, taking
turns to watch over the grill. This left the womenfolk free to sit and
chat, which we did, happy in the knowledge that no one was slaving
away behind the scenes in a hot kitchen. Instead of hamburgers, steaks,
and chicken thighs, I discovered what a barbecue was really for—cooking *salsiccia*. Having been brought up on English sausages, I was amazed

that these great coils of pork sizzling away over the coals went by the name of sausage. And why did it taste so good? The magic ingredient is *finochietto*—wild fennel, indigenous to Sicily, difficult to obtain elsewhere, and the quality of the pork, of course. But as Pilli explained, the pork for salsiccia must be chopped by hand, not minced. Woe betide any sausage that had come within inches of a mincing machine, and given the exacting standards against which any salsiccia would be judged, sausage buying was a controversial topic. There was an ongoing dispute as to which town in the area produced the best. Some claimed that you could buy perfectly good salsiccia in Mazara, others championed towns in the vicinity, but the issue was finally decided when Pilli came down in favor of Santa Ninfa, a bleak hilltop town rebuilt after a major earthquake in 1968.

The main alternative for the barbecue was fish, for which Mazara del Vallo is famous as the biggest fishing port in Italy. The first time I tasted fresh tuna, "the chicken of the sea" as Maria called it, I realized it was not so much about flavor as texture—firm but yielding to the bite. And how could I have lived for over a quarter of a century without coming across the queen of Sicilian fish—*pesce spada*—swordfish? Thick two-pronged slices, bearing an uncanny resemblance to a pulled tooth, were grilled over coals for just minutes until they were seared golden brown on the surface, remaining as white as snow inside. Also new to me were *triglie*, plump little red mullet, which were really a metallic orange rather than red, served with sprigs of fresh rosemary. In my pre-Sicilian life, squid had never held particular appeal, but then I had never had *calamaretti*—baby ones, puffed out with fresh breadcrumbs, parsley, garlic, and pine nuts. Best of all were *gamberoni*, the brick-red giant prawns, not as big as the tiger prawns of the Pacific, but to me, infinitely more flavorsome. Pasta, salsiccia, grilled fish and seafood, followed by huge grinning slices of watermelon, luscious peaches, and apricots from the garden were all washed down with robust local wines from Zio Giovanni's winery. Why was it all so delectable? Because of the quality of the basic ingredients? Or because every meal was a work of art, an offering of love by proud cooks in a beautiful courtyard on a perfumed Sicilian night?

Every night was party night at Santa Maria that summer. The mood was festive, and the Manzo family seemed to be in a continual state of euphoria. With Maria running the PR campaign, there was an open invitation to all members of the extended Family, friends, and friends of those friends to join us for dinner. Often we were catering for twenty to thirty people a night. After siesta and visiting hours, the scene of action shifted downstairs to the courtyard, where preparations for dinner were under way. With Antonietta and Antoine both off duty, dinner was an all hands on deck effort. While the women took care of the pasta, the men would organize the barbecue. I was given mundane tasks to perform, washing or chopping vegetables, setting tables, or, my speciality, keeping the children amused and out of their parents' way.

With too many to sit at one table, we would set up makeshift trestle tables in the courtyard. The children were not expected to sit. They perched on the staircase or plonked themselves on the lap of any available adult. Feeding the little ones was a complex task, which Maria Antonietta shared with her sister, Diana. Omar, an enchanting three-year-old, was the most difficult to feed. He would race round the courtyard chased by mother or aunt armed with a fork and a plate of pasta. But he was never hungry and regarded the flight from food as just another way of having fun. Bits of pasta fell to the ground to be polished off in an instant by Greta, the aristocratic hound. Night after night, as I watched preparations for dinner, I was puzzled at the sheer quantity of food prepared, especially as much of it was left over at the end of the meal. Coming from a family raised on the waste-not-want-not, eat-your-cabbage-or-you-won't-get-any-pudding principle, I wondered at the number of dishes, the huge amounts served at every meal, the carefree extravagance, all of it a smack in the face to penny-pinching English ways. I noticed that surplus food didn't get carefully stored away in the fridge the minute the table was cleared. Sometimes it was left out all night and thrown away the next day. That couldn't be right, surely. A horrible little voice in the back of my mind was telling me to think of the starving millions. If this family was trying to make economies anywhere, it was certainly not in the food department.

From the air of festivity that pervaded Santa Maria that summer, it was easy to forget that the Family was beset by financial difficulties. It was only when I heard the word *cambiale*—promissory note—again and again in family reunions that I gathered that the problem was debt. It turned out that Maria had still had cambiali that she could never realistically expect to repay. Typically she was undaunted. If she had to pay back a cambiale or two, so what? She could always take out another. There were plenty of people who would still give her credit. As a novice to the Sicilian way of life, I found it hard to deal with this concept of living in debt. And the Manzo-Vaccara debts were not the ones I was used to—manageable, socially acceptable mortgages repayable over a lifetime. No, these were frightening contracts, signed in a fit of optimism, which obliged the borrower to repay a sum far higher than he had actually received. Although I was not invited to the Family meetings that took place behind closed doors that summer, in due time I learned the solution to the looming crisis: Marcello would "invest" in Santa Maria by repaying the most threatening of the creditors. This way, Maria would not be forced to sell a large slice of land and the remaining debts could be rescheduled. Enthusiastic at this ingenious idea that would leave Santa Maria intact, the whole family heartily approved of our "investment." It didn't seem much of an investment to me at the time, but who was I to raise objections? Marcello had done his duty toward his parents, while his brothers congratulated themselves on the innovative business solution, *tranquilli* that Santa Maria's future was secure. In short, everyone was happy. Maria was not just happy, she was exultant. With her arms around Marcello and me, she announced to the assembled Family,

"*Avete visto*? Didn't I tell you something would turn up?"

MARIA

SEVEN

Maria

LIVING IN THE DEPTHS OF THE SICILIAN COUNTRYSIDE HAS ITS drawbacks, above all the constant need to *scendere*—go down to town and *salire*—come back up again. When Maria was in her prime, she gladly did the going down and coming up several times a day—to fetch Antonietta or take her home, buy two or three newspapers and large quantities of bread, pick up medicine from the pharmacy, and drop in on the Zii and Zie. Those were the pretexts, but in reality every trip to town had its networking objectives. The baker was an important source of local gossip, the pharmacist advised on Goffredo's ailments, and every time Maria strolled across the main piazza or down the smart shopping street of Corso Garibaldi, she would encounter people she knew.

Everyone she met as she went about town recognized her, for the Signora Maria was a *personaggio*—a great personality—well known and well loved in Mazara. She knew not just the names of everyone who had ever come into her orbit, but those of their children and grandchildren. She visited the homes of ex-gardeners, caretakers, or nannies, whose links with Santa Maria were long extinct, always remembering to take packets of sweets or biscuits for the children. She reminisced, exchanged anecdotes, and invited everyone to Get Close.

No visit to Mazara was complete without the obligatory stop off at the *circolo*, or club, the hub of social life in Mazara for men of leisure.

The *circolo* of the *nobili*, to which the Manzos and Vaccaras belonged, was not to be confused with the *circolo* of the *lavoratori* on the other side of Piazza Mokarta. As none of the members of the former had titles (with the exception of the Conte Burgios), and few of the latter had jobs, it seemed to me that the segregation of the two groups of men behaving idly represented aspirations rather than any real class distinction. The main salon of the circolo dei nobili with its dusty chandeliers and stained mirrors had clearly seen better days, as had many of its members, who were mostly elderly and down at heel. Men who spent the best part of the day there from Monday to Saturday would wander in and greet their fellow members with an enthusiasm that suggested long separation, before sinking into shabby leather armchairs to read the *Giornale di Sicilia* and chain smoke. Some would adjourn to the *salone* to play cards. While ashtrays filled up with cigarette stubs, elderly waiters would ply the salon to and fro, serving espressos or *zammu*— iced water flavored with aniseed, a summer drink reminiscent of Greek ouzo, dear to the heart of the Sicilian male.

Women could be honorary members of the circolo, but Maria seemed to be the only signora in town to exploit this privilege to the full. Whether or not accompanied by Goffredo or Silvio, her networking routine involved daily invasion of what was in practice a male enclave. Otherwise female presence was rare at the circolo, women only appearing to summon their menfolk for meals. The first time I was taken there my appearance caused quite a stir. Although we had interrupted their game, the group rose to their feet to greet Maria and make the acquaintance of the *straniera*. What beautiful manners, I thought, as they bowed and kissed our hands. Baffled at the sight of adult males engaged in a midmorning, midweek game of poker, I asked Goffredo and Maria about the gambling. Was this a normal way to pass the time of day? Didn't their friends have anything else to do? If they didn't have jobs, what did they live on? My hosts were vague on this matter.

"They all have land, and they probably just sell off bits of it when they need money," was their explanation.

The circolo dei nobili has closed its doors for the last time, and Maria, now in her mideighties, is so frail and her mind so confused that she is confined to Santa Maria. Although she has not been seen in Mazara for several years, thanks to a strong family likeness, Silvio, Marcello, and above all Pilli are instantly recognized as her sons. In summer, hardly a day goes by without someone stopping one of them to ask,

"Aren't you Signora Maria's son?"

Acquaintances are shocked to hear of her mental decline and shopkeepers even refuse payment when they recognize one of the three.

"What an outstanding woman she was. What a personaggio!"

She was indeed an outstanding personality, larger than life, with a history to match. At times she seemed more like a fictitious character than a real person, and I often felt that she could have stepped out of a work of literature, one in which the heroine is shaken from her pedestal of privilege and put to the test in one crisis after another. As a young woman Maria had had everything. She was rich, beautiful, accomplished, and adored by everyone, but what distinguished her from other women from a similar background was strength of character enhanced by a boundless optimism. It was to stand her in good stead when things began to go wrong. Her upbringing had hardly prepared her for the hard times that were to befall her in later life. A little rich girl (and Maria, the youngest of five, was Daddy's favorite), a typical *figlia di papa*, she would shop in the most expensive boutiques of Florence, Rome, or Palermo without ever bothering to ask the price.

"Just put it on the Vaccara account" she would say casually in *Pretty Woman* mode. Having been brought up in poverty himself, Luigi Vaccara was determined to give his children everything money could buy, above all the education that would elevate them to a higher social class. He sent his sons to northern universities to be educated as *signori* and to acquire the title of *Dottore* that he had never had, while Maria went to the Santissima Annunziata del Poggio Imperiale, Italy's most exclusive educational establishment for young ladies. When the five children grew up, married, and had children of their own, three generations of Vaccaras lived as one happy family in the grand home Luigi had built in Mazara adjacent to the wine factory. So grand was it by local standards that the Mazaresi dubbed it "the Vatican." While her four brothers assisted their father in various branches of the business with various degrees of enthusiasm, Maria handled the PR side, a role that she reveled in, even though it meant she often had to travel abroad with her father, leaving Goffredo at home and her baby sons in the care of nannies. The role of personal assistant to an industrialist with clients all over Europe and factories in Tunisia and Libya was a pioneer one for a young Sicilian woman in an age when her peers rarely strayed beyond the confines of home. Young ladies in Sicily in the 1930s definitely did not get mixed up in *affari degli uomini*—men's affairs—and certainly never left their husbands and children behind to go on business trips.

When Luigi died unexpectedly just after Marcello was born, the Happy Family was broken up. Unfortunately he had left his heirs unprepared for the future. His four sons had been brought up to be gentlemen with an aptitude for spending money rather than making it, and Maria, although active in maintaining relations with clients, had never been involved in the administration. It is doubtful whether she had ever even contemplated the idea of a budget, let alone had to meet one. Only Zio Giovanni had inherited Luigi's flair for business, and if he did not actually expand his position, he did manage to hold on to most of his inheritance. For the rest it took just one generation of extravagance and misguided investments to squander the Vaccara fortune. The destruction was complete before Luigi's grandchildren even reached adulthood.

Luigi Vaccara died intestate. While the lion's share of his industrial empire was divided among his four sons, Maria inherited various pieces of real estate, including Villa Santa Maria and the surrounding vineyards. Although she must have realized she had received the thin end of the inheritance wedge, in the interest of Family Peace she chose not to contest her brothers' division of the spoils.

Nevertheless, not content to administer her assets and live off the income, she decided to embark on an industrial enterprise of her own. I suspect there was a subconscious desire to outstrip her brothers, who had inherited more lucrative slices of the pie, and to show to the world she was her father's daughter. In the 1960s she launched her own wine label at Santa Maria, converted the warehouses, stables, and workers' accommodation into a vast modern wine factory, erected mammoth-sized concrete vats, and installed all the latest machinery. To carry out the conversion project, she had to take on substantial debt, which she optimistically reckoned would be paid off with the proceeds of the business. Borne on a wave of enthusiasm, she went on an industrial spending spree. With Goffredo helpless to stop her, the debts began to mount. Lack of business acumen combined with irrepressible optimism doomed the project from the start. The outlay was far too great for the volume of production, and, despite Maria's energetic promotional activities, sales were disappointing. Eventually the banks called in the loans, and the Industria Maria Vaccara went into liquidation. Santa Maria was rudely stripped of decorative emblems and statuary and the contents auctioned. Everything the villa contained, from antiques and paintings to Goffredo's antique gun collection, even Maria's furs were handed over to satisfy the creditors.

Meanwhile, Maria met the threat of bankruptcy with the same enthusiasm that had led her to overspend in the first place. She embarked on all kinds of fund-raising efforts, from becoming a real estate agent to writing novels. She begged and borrowed from friends and family, and if she did not actually go as far as robbing a bank, she certainly deployed her charm and beauty into seducing more than one local bank manager into signing generous promissory notes on her behalf.

In literature and the media Sicilians are portrayed as people pas-sively resigned to their unenviable lot, victim of centuries of foreign invaders and compliant with the Mafia. Maria could not be further from this stereotype. No one could accuse her of pessimism or passive resignation. In fact she was very un-Sicilian in her belief that in loss and misfortune "things would readjust themselves." At times she would expound to me at length about the treasures she had had to sell to pay off the debts, particularly the works of art her father had acquired.

"We had paintings by Guttuso and Consagra." At first the names meant nothing to me, but later I learned that Renato Guttuso, who came from Palermo, was Italy's most celebrated contemporary painter and Consagra, a distinguished sculptor from Mazara.

"Yes my father was Consagra's first patron. He commissioned a marble bust of himself. It was the first piece of work Consagra ever sold. He stayed here at Santa Maria for many months with his American wife, and they both made paintings of the villa."

"What happened to those?"

"Like everything else, they were all sold."

"Who bought all these things?"

"They went to auction, mostly. A lot of them ended up in the homes of relatives."

"Weren't you sad about that?"

"Yes, but at least I know where they are, and one day, when things look up [an expression that would become very familiar], I'll be able to buy them back."

Not only did she always hold tightly to the conviction that things would look up one day, but she never felt sorry for herself or bemoaned the fact that she had come down in the world. She once asked me which I thought was better—to be poor in one's youth and rich in old age or the other way around. Whereas most people would think it obvious to be poor first, rich later, on the grounds that one doesn't miss what one has never had, Maria could not imagine herself without her memories of a privileged and gilded youth and the question she had put to herself was a genuine dilemma.

Second only to her optimism, Maria's most outstanding quality was her sociability. Given that Italians are more naturally sociable than the British, and Sicilians, by the Goethean principle, the most sociable of

Italians, Maria still managed to put her countrymen and women in the shade. It was impossible not to be swept off your feet by Maria. She dominated the Family and any social gathering she attended. As Marcello's bride-to-be, overwhelmed by all the Getting Close, I had assumed that the intensity of Sicilian social life would be restricted to special occasions such as the lead-up to a wedding. This may have been so with other families, but, as I learned on subsequent visits to Sicily, it did not apply to ours. Maria liked to be surrounded by people, and her ideal scenario was open house at Santa Maria. If people couldn't come to us, we would simply go to them. Getting Close went on day and night on every summer visit, year after year. The worst thing Maria could conceive of was to be left alone "lonely as a dog," a fate that she herself made sure never befell her even for a moment. She was a tireless hostess, and both she and Goffredo reveled in the company of young people, who in those days were Marcello's generation, the cousins, and the younger echelons of Honorary Family, all of whom she addressed as *voialtri*—you guys—and all of whom adored her.

By the time I met her she was already in straitened circumstances, but the self-confidence of her youth was undiminished. At the exclusive school in Firenze she had rubbed shoulders with the daughters of some of the most famous names in Italy—Marconis, Pirellis, Agnellis, and Mussolini's daughter, Edda, not to mention the scions of the aristocracy. When Maria recounted tales of boarding school life in the 1920s, I couldn't help thinking that in such a social and political hothouse, a little girl from a small town in Sicily might have felt out of place. But I had underestimated the huge supply of self-confidence she acquired long before self-esteem came into fashion. Later, when I got to know her better, Maria admitted that in fact she had been bullied by some of her snooty dorm mates for her provincial background and, especially at the beginning, for her Sicilian accent. Once during a trip to Florence, I made a detour to the Poggio Imperiale to take a look at the place where Maria spent twelve years of childhood and adolescence. Still run by nuns, it has shed all aristocratic connections and is now a state school. The name means the Royal Lodge. Originally a hunting lodge for the Medici Dukes of Tuscany, the magnificent baroque decoration is still in place. Frescoes of mythological subjects and gilded columns abound, even in the dormitories. As I stood there admiring the ceilings, I tried to imagine Maria as a six-year-old engaged in a pillow fight, pink-faced and flustered, thumping away at her bosom pal, Gioia Marconi (daughter of the Father of Radio), or in a sadder scenario, lying awake in the dorm, unable to sleep for the snoring of blue-blooded comrades, her gaze forcibly directed at the sylvan scenes on the ceiling.

If the tough little Sicilian kid stuck it out in that particular ambience, she did so by sheer force of character. When Marcello was a small child, he realized that his mother was different from other women. Maria had always loved to be behind the wheel of a car, and in fact she was the first woman to get a driver's license in Mazara way back in the 1930s. In those days the Sicilian countryside had more mule tracks than roads, and you could count the number of motor vehicles in Mazara on two hands. A woman driving a car was in the category of an E.T. landing. Marcello, as an under-ten who did not go to school and had no contact with children other than his brothers and the Vaccara cousins, was reduced to an extreme state of mortification every time Maria took her sons out in the family Lancia. They would be pursued along the alleyways of Mazara by the street children of the port, shrieking at the tops of their voices,

"*Talia, talia, una fimina chi mania!*—look, look, a woman driving!"

Maria liked speed and, not content with cars, she bought a motorbike, thus making another record as the first woman to ride this macho vehicle in the province of Trapani, perhaps in all of Sicily. There is a wartime photograph in the family album of a beaming Maria in khaki, apparently driving a tank, while military personnel look on admiringly.

"So Maria, you must have been not just the first, but the *only* woman in Sicily to drive a tank," I exclaimed when she showed me this picture.

"Oh, I didn't actually drive it."

I refrained from asking why the uniform, but by then I was familiar with the family penchant for dressing up. When not engaging in masculine pursuits, she kept up her fluent French, played the piano at family soirées, and kept fit by swimming for miles in the Sicilian Channel. With her film star looks, millionairess wardrobe, and Florentine chic, in provincial Sicily she shone like a star.

At the Poggio Imperiale, Maria had acquired a wide circle of rich and important friends, most of whom were based in north and central Italy. Frequently invited to the villas and palazzi of these families, she was at home in the sophisticated world beyond Sicily, but her attachment to her family, particularly her father, was so strong that it bordered on worship, and it never occurred to her that it would be better to live anywhere else in the world than Sicily.

"Che bella è la Sicilia" was a sentiment she would repeat again and again, just like Raffaella two generations later. Maria's sicilitudine included not just the Family but Santa Maria, Mazara, and a passion for the basics of Sicilian life, such as pasta with salsa and Sicilian bread.

Although the intensity of her social life and frequent travel made Maria an unconventional mother by the standards of her day, at the same time she was very much the dutiful Sicilian daughter. When her father, Luigi, died, his widow, Francesca, went to live with Maria, her only daughter, who looked after her for twenty years. I never met Marcello's grandmother, but from what he has told me, she was home loving and simple but wise, the exact opposite of her worldly, glamorous daughter. Born in the 1880s, she had grown up in an age when the children of contadini helped their parents in the fields and vineyards instead of going to school (Luigi was an exception and even he left school at twelve), and most girls reached adulthood unable to read or write. As teenagers, Marcello and his brothers would take turns keeping their grandmother company in the evening when Maria and Goffredo were out, as they so often were. Despite her grandsons' attempts to teach her to read on these occasions, she never managed to get past the first few letters of the alphabet, and in the end the literacy campaign was abandoned. Not that Nonna Francesca minded. She had been content to sit at home while her husband pursued his career and the social life that went with it. After his death she donned the top-to-toe black of the Sicilian widow. Although always at the center of family gatherings, she declined to attend her daughter's elegant soirées, preferring to retire to the kitchen, where she would receive her contadini relatives. Typically of the illiterate, Nonna Francesca had a good memory, and it was she who taught her grandchildren Sicilian fairy tales and folk songs. Sagacious in her simplicity, she had a proverb for each occasion. Once when Maria and Goffredo were opening bottles of champagne to celebrate the sale of a parcel of land, Nonna Francesca cryptically reminded them "*Cu vinni, svinni*—he who sells, loses out." She did not need a fancy education to tell her that the sale of a vineyard to pay off a debt was not something to celebrate.

No sooner had we met than Maria engaged me as her literary agent. I never did manage to get *The Dons* published in Italy, although I tried my best. Secretly I had my own misgivings about the book. How much did Maria really know about the Mafia, I wondered. Hadn't she been whisked away from the island at the tender age of six? And wasn't the Family particularly reticent on the subject? After taking the manuscript to three major publishing houses and receiving rejections from all three, I gave up. Maria did not hold my failure against me at all. By the time I returned to Sicily, I was Marcello's fidanzata, and in the excitement of the impending wedding, she seemed to have forgotten that she had ever entrusted me with the mission. It was clear that she was no longer writing, and I wondered, with her busy social schedule, how she had ever managed to fit in the long hours required to write six novels in the first place.

Whatever her ability as a writer, Maria was certainly a great reader. She was passionate and knowledgeable about Sicilian literature. One of her first questions to me was "Have you read *Il Gattopardo*?" I had to confess to only having seen the movie of *The Leopard*, whereupon she

gave me her copy to read, saying, "Read it if you want to understand Sicily." And so I read Tomasi di Lampedusa's great historical novel, in which a Sicilian prince meditates on the imminent extinction of the old aristocratic way of life under the Bourbons, against the background of the great political upheaval of the Risorgimento. Whereas Visconti's film version was all glamour and opulence, the book itself is a profound exploration of the island and its people. "The Leopard" is Don Fabrizio, the Prince of Salina, who represents the old order. The new political class that replaces the old is personified in Don Calogero Sedàra, nicknamed "the Jackal," the nouveau riche upstart, "all munching and grease stains," "a little heap of cunning, ill cut clothes, money and ignorance." Married to the illiterate daughter of a Mafia hit man, he has a beautiful daughter, Angelica, who captivates Tancredi, Don Fabrizio's penniless ward and nephew. Angelica is elevated out of the peasant class of her parents and presented to the world as ultimately desirable for her beauty and her fortune. To further his nephew's career and satisfy his passion, Don Fabrizio has to court Sedarà, the man who stands for everything he despises. For an outsider looking for an insight into Sicily and its history, the musings of the disillusioned Prince give a pessimistic view. With Don Fabrizio as his mouthpiece, the author argues that Sicilians have been corrupted by thousands of years of colonial rule:

> All those rulers who landed by main force from every direction, who were at once obeyed, soon detested, and always misunderstood, their only expressions works of art we couldn't understand and taxes which we understood too well and which they spent elsewhere: all these things have formed our character, which is thus conditioned by events outside our control as well as a terrifying insularity of mind.

As I read the book, I couldn't help noticing similarities between Maria and Angelica, the heroine—the parents' humble origins, the father's rise from rags to riches (but not the Mafia connection), the transformation from country lass to polished beautiful young woman. Even the school in Florence that accomplished this was the same famous establishment that Maria had attended. If at first I was baffled at Maria's enthusiasm for a book that expressed such scathing views of her beloved Sicily, gradually it dawned on me that she recognized herself in the role of the main female protagonist. *The Leopard* was her favorite book, not just because it was beautifully written, but above all because she identified with the heroine.

MARIA'S *PREPOTENZA* TRAVELED LONG DISTANCE. "*BACIONI alla piccola* Maria—big kisses for little Maria"—was on the telegram congratulating us on the birth of our baby daughter. No one had told Maria that the baby would be named after her, but she had automatically made that assumption. I had other ideas, however. Torn between Melissa, a character in Lawrence Durrell's *Alexandria Quartet,* and

Angelica, the heroine of *The Leopard*, I thought I would have one or the other, or both. Maria was a perfectly respectable name, apart from the West Side Story associations ("I've just met a girl called Mareeeeah") but having gone to all the trouble of having the baby, I claimed the privilege of choosing her name. Marcello agreed with me, and bravely he braced himself to break with tradition—that is, until the telegram arrived. He was then in the unenviable position of choosing between a bossy mother and a truculent wife. In the end *truculenza* won the battle against prepotenza and we placated Maria with the promise that her name would appear in third place. Our elder daughter goes by the name of Melissa Angelica, and fortunately my mother-in-law has never asked to see her birth certificate.

As long as she had Goffredo by her side, Maria was able to face the vicissitudes of life with equanimity. His unexpected death, when they were both still in their sixties, was the catalyst precipitating a change in her personality that gradually developed into mental illness. Distraught though she was, Maria gained comfort from the complex rituals involved in the Sicilian way of death. They seemed to provide her with the displacement activity she desperately needed to survive the crisis. With the help of her favorite niece, Franzina, and Antonietta, she took care of the laying out of the body, the washing and dressing, a last labor of love for Goffredo and not for the hands of strangers in an anonymous funeral parlor. Then she had to attend to the wake, the decoration of the *portone* with purple swags, the visitors' book, the announcements, and the preparations for the funeral. From morning to night the Family, Honorary Family, ex-employees, and acquaintances came to offer their *condolanze*, a ritual that went on for nine days according to Sicilian custom. When the ritual mourning period was over and it was time for us to leave Santa Maria, we did so with heavy hearts, burdened with guilt. For a Sicilian family in the 1970s it was taken for granted that the children would look after their parents. Maria, who had never spent a single hour by herself in her whole life, certainly could not be expected to live alone now.

Care of a widowed mother would normally have fallen to daughters, but as she had none, to her daughters-in-law. Unfortunately for Maria, both her married sons had wanderlust. Maria Antonietta and the children had followed Pilli to a new job in Zimbabwe, and Marcello and I had to go back to Brazil. So the burden of filial duty fell on Silvio, the youngest son, who had no choice but to step into his father's shoes. At this stage no one, least of all Silvio, was fully aware of just how great that responsibility would be.

At first, with the support of the Family and a large network of friends, Maria coped with her bereavement admirably. An old school-friend invited her to become president of the local branch of a national women's organization, a role that involved traveling, speech making, and a great deal of networking. Although this occupied most of the daylight hours, for the first time in her life her evenings were empty. She was now dependent on Silvio's company. A bachelor in his thirties should be entitled to his own social life, but if Silvio wanted to go out at night, first he had to organize *la mamma*. Friends and family were understanding. The Signora Angelo, Marcello's former governess, Franzina, and other family members offered to keep Maria company in the evening. In the end the Family devised a sort of dumping system whereby Silvio and Maria dined together at Santa Maria, after which he would escort her to the home of a willing relative or friend. This would allow him a few hours of freedom before he had to go back to pick his mother up. Although Maria was perfectly happy to socialize until the small hours every night of the week, her hosts' gregarious instincts were usually less developed than her own, a fact she was unlikely to notice. Silvio had to make sure he was back in time before the welcome wore off. He had the misfortune to be unmarried, with a mother so very attached and dependent, and yet so very sociable. With this limitation on his social life the prospects of renouncing bachelor status did not seem good.

Soon after Goffredo's death and our departure from Sicily, we invited Maria to come and stay with us in São Paulo. At first she was delighted with everything in Brazil, out little chalet-style house with its tropical garden, the shopping malls, then unknown in Italy, the smart restaurants, and lush parks. She doted on her baby granddaughter, whose antics inspired her grandmother to call her Eleonora Duse, after the famous actress of the Belle Epoque. She was thrilled to find we had a circle of Italian friends, who issued dinner invitations in her honor every night. Then when the invitations ran out, she took it on herself to invite everyone back. Within a fortnight however, the novelty of it all had worn off. Maria was homesick. Unable to sit still for a minute, she would pace up and down, worrying that Silvio could not manage without her. There was work to do in the garden, payments to make, Antonietta and Antoine to supervise.

"What's wrong?" we asked. "Surely you don't want to go home yet? Your ticket is for a month's stay."

"Haven't you read the newspapers? That swine Gaddafi is threatening to attack Italy. Libya's only a stone's throw from Sicily, you know. If there's going to be a war, I've got to get back. For all I know, he's got his missiles pointed straight at Santa Maria."

There was no reasoning with someone so used to getting her own way, and we had to organize her flight back to Italy two weeks ahead of schedule with the escort of a reluctant Italian expatriate.

When we returned to Sicily the following summer, she had forgotten her preoccupation with Gaddafi and his missiles. The latest crisis was the arrival of the royal yacht *Britannia* at Palermo. If only she had known about this in time, she would have invited Queen Elizabeth to Santa Maria. What a *brutta figura*—bad impression—she would make, if the Regina Elisabetta got to hear that Maria Vaccara had shown discourtesy by not issuing an invitation to Santa Maria. Now it would be too late, or would it? At first we thought she was joking, but no, she was absolutely serious, insisting that it was the least she could do to extend the appropriate courtesy to the Queen of England. It took the powers of persuasion of all three sons to stop her telephoning the mayor of Palermo.

If we had been more prescient, we would have recognized these eccentricities as a warning of the mental decline that was to come. But how were we to identify it? Those close to her were used to and tolerant of her whims. She was loving to her family, charming to her friends, and kind to everyone. If she was just a little prepotente and stravagante occasionally, one could attribute it to her bereavement

Since we saw Maria only once a year during this period, the change of personality was more perceptible to us than to those who saw her daily. The first thing we noticed on our next visit was the deterioration of the garden. She had completely lost interest, and with no gardener employed to water and weed, it was turning into a brown, desolate waste. Apart from the garden, she gradually dropped other interests once so dear—literature, local charities, and the women's association. Her fondness for company was developing into an obsessive fear of being alone. As long as she could drive her battered Cinquecento, she was independent in the daylight hours. She rarely lacked an excuse to scendere, as there was always the bread to fetch twice a day, the pharmacy to visit, and the circolo to check out. Sometimes we would find four or five newspapers she had picked up on trips to town, unopened by the end of the day.

Visitors to Santa Maria were far less frequent than before, however, and the invitations to dine out were now few and far between. No doubt the regular "dumping" sessions had worn local hospitality thin. Nevertheless, even when Pilli's visits coincided with ours, and there were six adults in the house as well as the children, this was not enough company for Maria. Staying home in the evening was absolutely out. If no one was expected after dinner, we would have to do some Getting Close ourselves. Or else we would drive out to the fishing village of Torretta, where we would be sure to meet friends in the piazza. As soon as supper was over, however late the hour, Maria was determined to go out. Every night she would ask the same question over and over again.

"*Dove andiamo sta sera?*—where are we going tonight?"

As far as I was concerned it was better to receive visits than to make them. I was an inexperienced mother, but instinctively I knew that cavorting into the small hours was not good for a small child. I had ideas of circumventing the late night Getting Close by putting the two-year-old Melissa to bed at a reasonable hour and slipping away myself around midnight with the excuse of having to rise early for the baby. Having resigned myself to the fact that my mother-in-law would never babysit for us—it would never have entered the waiting room of her mind—I resolved to assert myself only if Maria's whims interfered with the baby's well-being. On one occasion when Melissa was already asleep in bed and the family decided to go out, I bravely announced that I would forgo the outing and stay home with Melissa. Somehow I had stepped into a minefield. Marcello would not leave his wife and baby daughter alone in the villa, and Maria would not go out without Marcello. Maria Antonietta and Maria told me to get Melissa up. They couldn't understand why I had put her to bed in the first place. Why didn't I just let her stay up and party? Why was I punishing her? A debate ensued which Marcello made sure he stayed well out of, and which cast me in the role of spoilsport-cum-child-abuser. In the end Maria took the problem into her own hands—literally. She left the room, only to return triumphant minutes later with a sleeping baby in her arms. Plonking Melissa down, in nappy and nightie in front of the assembled family, she declared in triumph,

"She was awake already. *Andiamo?*"

This was the nearest I ever got to a clash with Maria. Fortunately she was so good-natured that it would never have occurred to her that I might take exception to her interference, and I never stayed long enough in Sicily for any real friction to develop. In fact she often took me aside to tell me I was her favorite daughter-in-law, an honor I had certainly done nothing to deserve, and which was hardly fair on Maria Antonietta, considering she had never caused Maria any trouble, apart from making off with her eldest and favorite son. In moments of fondness Maria would call me to her bedroom and press little gifts into my hands, not jewelry, for anything of value had been sold long before, but fripperies she had conserved from the fifties: brooches in fake coral and turquoise, mantilla combs, a tortoiseshell cigarette case.

"But I don't smoke!" I protested.

"Well you can always put your lipstick in it."

She also pressed me to take clothes she had worn in a more glamorous era, hand-painted fans, a pink damask stole, brocade jackets, and gathered skirts with impossibly small waists. Did she really think I could go out at night dressed like a 1950s Sicilian heiress? Of course I couldn't, not even at the Rio Carnival. I also balked at the thought of how much of our luggage allowance all this gear would take up. We couldn't leave it behind, Marcello insisted. We had to take it. We duly packed enormous amounts of ridiculous and useless clothes into our suitcases, the only upside of the business being a regular supply for the Flowers' dressing-up box.

In the mideighties Silvio became engaged to be married. His bride was the daughter of an elderly Sicilian gentleman who had emigrated to San Diego in his youth, where he set up a chain of barber shops, married Anna, the daughter of Sicilan immigrants, and with her raised a family. When Joe reached retirement and decided to return to his home town in Sicily, he brought his wife and two unmarried daughters with him. To give up their jobs and say good-bye to family and friends, not to mention the lifestyle of California in exchange for village life in Sicily, must have been a traumatic experience for the two girls. Jobs would not be easy to find, as Nanette and Trina did not speak Italian—only English and Sicilian dialect. Already in their thirties, they were well past marriageable age by Sicilian standards. When Joe engaged Silvio as the architect to build a house for him in neighboring Campobello, the architect fell for the tall blonde elder daughter, who spoke Sicilian with an American accent, and the courtship was under way.

With the new addition to the family, Silvio was not just gaining a wife but Maria was gaining a long-awaited resident daughter. It must have been from a Sicilian sense of family duty that Nanette agreed to move into her mother-in-law's establishment after the wedding, the same sense of family duty that brought her to Sicily in the first place. It seemed the perfect solution, and everybody was happy with it, especially Maria, who, from then on, would never have to worry about being alone. Then there was the excitement of the wedding to organize. The dilapidation of Santa Maria was too far gone by then to contemplate a ceremony in the chapel, but at least this time the wedding would be done properly. Silvio and Nanette were married in the Cathedral of Mazara, with the bride in white and the ceremony followed by a wedding feast at a local hotel. There was one small hitch however, at least as far as Maria was concerned—the honeymoon. The newlyweds were only going as far as Paris, but they would be away for two whole weeks, and she found the prospect of separation from her youngest son very hard to bear. Even though Silvio had arranged for a cousin to stay at Santa Maria to be with his mother, she was in despair. Unable to persuade them to forgo the honeymoon, she telephoned the hotel in Paris night and day, desperately trying to convince them to cut the trip short and come back to her and Santa Maria.

But she survived, and Santa Maria took on a new lease of life. An earthquake in 1981, which took no lives in Mazara, inflicted structural damage to many old buildings, including Santa Maria. Eventually the comune came up with financial contributions for the buildings affected. This paid for repairs to wall and roof, plastering, painting, and rewiring, while Nanette's parents made a generous wedding present of new bathrooms and kitchen so that the living quarters on the first floor acquired a new look. The Sicilian side of Nanette was happy to take on her mother-in-law on a permanent basis, but since she also came from California, there was no way she would put up with plaster falling off walls, dangerously exposed wiring, and smelly, hydraulically-challenged bathrooms.

With the villa itself no longer about to fall down, and Maria over the moon with a resident daughter-in-law in place, we could all breathe a sigh of relief. From our base in Jakarta and later in Rio de

Janeiro, we continued to visit once a year. But with the addition of Clemency to the family, we felt that four of us were too many for extended visits. When Silvio and Nanette had children of their own, space became a real problem. Only the first floor was habitable, as the earthquake funds had not stretched to renovating Marcello's share of the villa—the ground floor. Every time we came to Sicily, Maria implored us to stay indefinitely, but we felt sorry for Nanette, who clearly had her hands full. As well as her own two small children, she now had to look after Maria. The last thing she needed was an extra family on a prolonged stay.

EVEN WHEN THE FLOWERS WERE SMALL CHILDREN THEY REAL- ized that there was something about their nonna that was not quite right. Every now and then conversation with Maria would move into the realm of the bizarre. She would regale us with tall stories—how she had just swum a hundred lengths in an Olympic pool.

"But Nonna, that's impossible," the Flowers would protest.

"*Ma certo, gioia mia*, you don't know what a good swimmer your nonna is. When I was a girl, I once swam all the way to Tunisia and back." Either Nonna was telling fibs or her mental faculties were seri- ously compromised. She would repeat herself constantly or ask the same questions over and over again. By this time she was aware that she was losing the battle with sanity and she would plead with us—*scusami, scusami*—when we answered the same question for the umpteenth time. When the memory loss and confusion precipitated, we were told she had Alzheimer's, the same form of senile dementia that had struck two of her brothers.

My mother-in-law had always had a radiant personality, charming everyone she met and loath to find fault in others. Alzheimer's changed all that. Whereas she had always been tactful about any family members she secretly disliked, she now let rip in front of the people concerned, informing them in no uncertain manner that they would no longer be welcome in her house. She also developed a fierce hatred of perfectly innocent people. One such victim was Nanette's mother, Anna, whom she accused of trying to kill her. Maria's mental state had deteriorated so badly that she no longer recognized herself in the mirror. Nanette went into her room one day to find her directing one of her venomous attacks on her own reflection. She had found a new demon to plague her, and one that would not tactfully retreat. Although still in good physical health and far more agile than I was (after a series of hip operations, I was on crutches for much of this time), she was in a state of mental confusion that caused havoc in the household. She became more and more difficult to live with, repeatedly demanding food, having forgotten that she had just eaten, removing clothes from wardrobes and spreading them on the floor all over the house, taking bread rolls and fruit from the kitchen and squirreling them away in hidden corners of the villa until a nasty smell or a trail of ants would lead to detection. Sometimes she would adamantly refuse to go to bed; on other occasions she would get up in the middle of the night to visit other members of the family who were fast asleep.

In spite of the sadness of Maria's predicament, occasionally it was hard to repress the instinct to laugh. Old habits die hard and she would still rush to answer the phone, even though she had no idea of whom she was talking to. One day we found her talking into a hairdryer, which she had mistaken for the telephone—the word for hairdryer in Italian being *fon*—shouting at the top of her voice in frustration that she was getting no response.

Before long she had difficulty recognizing anyone outside the close family circle. Family and friends would still come to Santa Maria to see her, but they would leave with the sad comment: "She didn't recognize me." Even Franzina and Antonietta, whom she saw every day, were strangers to her. We were shocked and saddened when we discovered that she no longer knew the Flowers or myself. Then it was Marcello's turn. One evening we took her out for an ice cream, and as Marcello handed over the cone, Maria whispered in my ear:

"*Che bravo ragazzo*—What a nice boy. Do tell me his name."

We all dreaded the next stage of Alzheimer's, the physical dependence, when she would have to be bathed, clothed, and fed. Strangely enough, when it came, it was almost a relief. The aggression slowly ebbed, as she relapsed into infancy and responded to those around her with gestures of affection. By now she knew no one but Silvio, Nanette, and their children, Massimo and Stefi. Although she couldn't communicate with language, she could make herself understood with sounds and gestures. It was clear that although she could not identify us, she knew we were close to her.

In Britain most people in Maria's condition would have been shunted off to institutional care long ago, but in Sicily not only are the links between young and old much closer, but there is also a culture of shame attached to anyone unable to look after his own. Consequently, old people's homes are mainly for the destitute. Maria is exceptionally fortunate in having Nanette for a daughter-in-law. But even if she still had the faculty of reason, she would expect no less, for she had looked after her own mother until the end. She now spends her days in the upstairs sitting room, present but not participating in family life. Although she is no longer aware of what is going on around her, Silvio and Nanette are a constant presence, and both Marcello and Pilli come back to see her year after year. Looking at her frail form and her still fresh pink-and-white complexion, I wish I could give her a penny for her thoughts. But whatever passes through the clouded recesses of her mind, there is a certainty reflected in her smiles and gestures. She is surrounded by people she loves, and as she gazes out onto the pines and cypress trees that she planted half a century ago, there is no doubt in my mind that she knows she is at Santa Maria, loaded with memories of Goffredo and her father, Luigi—Santa Maria, where she always wanted to be.

GOFFREDO WITH PILLI AND MARCELLO

EIGHT

All Saints and All Souls

JUST MONTHS AFTER OUR DECISION TO GO FOR SANTA MARIA, on a foggy February morning, I step through the sliding doors into the arrivals hall at Linate airport, pushing a trolley laden with suitcases. During the spring half-term visit to London, in between slots of quality time with the Flowers, I have indulged in my annual book-shopping binge. Instead of fiction, travel and history, my usual fixes, this time the books come from a different department altogether. Fired with enthusiasm for the imminent renovation project in Sicily, I have splashed out on glossy tomes packed with inspirational photographs and titles such as *Converting Old Houses, The Home Decorator*, and *The Dry Garden*. I can't wait to unpack and show them to Marcello. Expecting the usual enthusiastic welcome, I am surprised to see him standing at the back of the crowd with a glum look on his face. I know immediately that something is wrong, but I can get nothing out of him until we get home. Then he blurts it out.

"*Mi hanno licenziato*—I've been fired." Used to Marcello playing the fool, at first I assume he is joking. When I realize that he isn't, I have to pinch myself to make sure I'm not dreaming. Disasters of this kind simply don't happen to us. Why? How? Whose decision was it? The question pour out as I try to hold back the tears.

"When did they tell you?"

"Last Friday."

"The day I left!" I had been cheerfully enjoying myself in London with the Flowers, shopping and lunching with friends while Marcello had borne this terrible blow—all alone—for a whole week.

"Why, why, why on earth didn't you tell me then?"

"I didn't want to spoil your holiday."

The sudden takeover of Marcello's firm by an American multinational is just the latest in a series of maneuvers between the giant pharmaceutical conglomerates. The frenetic episode of downsizing that ensued in this case has made him unemployed after twenty years in the same company. Our lives have been turned upside down almost overnight. Suddenly stripped of the financial security that Marcello's job had provided, the future looks bleak. At fifty-six, how is he going to find another job? His severance pay will get us through the next few months, and there is the modest nest egg earmarked for Santa Maria, but what then? Once over the initial shock and with Marcello calmer for having shared the problem, we sit down and thrash out a strategy. I will step up my teaching schedule while he looks for consultancy work. But there are far too many unemployed executives from the pharmaceutical industry competing in the market for fewer and fewer jobs, and as a language teacher, I can only earn a fraction of what Marcello was earning. There will have to be a tightening of belts, and we agree that the priority is the Flowers' education. As for Santa Maria, we have no choice but to shelve the renovation project. After all these years of debating the subject, and just when we had made up our minds to go ahead, the financial carpet has been rudely pulled from under our feet. We make a special trip to Mazara to break the news to Silvio and Nanette. Instead of poring over estimates to choose a contractor, as we had planned, we have to tell them we will only go ahead when Marcello finds work. We say "when" but, we all know that it could be "if."

What until recently was a plan to restore a castle in the sun is now nothing but a castle in the air. For the time being the ground floor of the villa will be left to the damp and the rats.

In the months that follow, Marcello keeps himself busy networking international contacts, but apart from one consultancy job, there is no change in the situation. By now August is approaching, the dreaded month when Milan transforms itself into the hottest city in the whole of Italy, with temperatures rising over 100 degrees Farenheit and humidity hitting 90 percent. Just to add to the general discomfort, the mosquitoes that breed in those canals so thoughtfully designed by Leonardo da Vinci, are trying our forbearance to the limit. Meanwhile the whole country is awaiting Ferragosto, the mid-August celebration of the Assumption of the Virgin Mary, which heralds a general downing of tools and putting up of feet. In northern land-locked cities, it also means mass exodus. For at least two weeks Milan becomes a ghost city. Factories and offices close, likewise many shops and restaurants. If you stay, we were told, you will have the city all to yourselves. The prospect of having Europe's most polluted city to ourselves for the hottest, most humid month of the year does nothing to lift our flagging spirits. We too will have to get out.

For the first time in twenty years we will not be making a summer visit to Santa Maria. Maria's illness is now so advanced that she requires round-the-clock nursing. It would be selfish to add to Nanette's burden by presenting her with four extra people in the house. Pilli's establishment is, as always, bursting at the seams. No other accommodation is available nearby, so we resign ourselves to going elsewhere. When my sister suggests a joint family holiday in Tuscany, we find a villa for rent on the road that winds round the back of Cortona. The rent is high, the villa uncomfortable and badly equipped, but the minuscule pool tucked into the side of the hill and the spectacular view over the Val di Chiana make up for the drawbacks. Italian summers are supposed to be hot and dry, but this year the weather lets us down. We have forgotten that in northern Italy August can also bring thunderstorms. As the clouds gather and shards of water bounce off the bijou swimming pool, we take refuge in the culture trail, doing our best to lap up all the art available, from Giotto in Assisi to Fra Angelico in Cortona, and Piero della Francesca everywhere. But we do so along with thousands of other tourists, darting to and fro across the valley in their hired cars, like little balls in one of those plastic hand-held mazes that come out of Christmas crackers. Whenever we alight from the car to contemplate a Renaissance masterpiece or to explore an Etruscan ruin, we find ourselves shoving and pushing our way through hoards of other punters doing exactly the same thing. We don't venture afar. To attempt Florence would be madness. Who wants to face Italy's most illustrious city of art, where the streets are constantly choked with a heaving mass of humanity? We soon realize that Tuscan and even Umbrian villages are tainted by globalization. The piazza of the nearest *borgo* on the shore of Lake Trasimene is taken over by overbearing four-wheel drives and their equally overbearing passengers, while the greengrocer displays yesterday's *Times* and the *Washington Post* alongside the potatoes. So much for our idea of getting away from it all.

By the end of the holiday the Stendhal syndrome of overexposure to culture has hit us all. The adults are longing for home comforts and the teenagers are definitely Tired of Tuscany. Clem says she never wants to see another fresco again in her life, and her sister and cousins probably share her opinion but are too polite to say so. Suffering from a surfeit of Florentine Madonnas and Etruscans in all their manifestations, by the Nine Gods I silently swear that from now on we will take our Renaissance culture off-season and in small doses. Meanwhile, fed up with the rented villa and the bad weather, missing her nonna, the *cugini* of Santa Maria, and the sea, Melissa voices the hitherto unspoken question.

"When will we be going back to Sicily again?

ONCE BACK IN MILAN I AM CAST IN THE ROLE OF PERSEPHONE returned to the Underworld. Although I am fortunate in that my consort is a comforting Marcello, not a scary Hades, once again I face the same old problems of everyday life. Nothing has shifted. Marcello resumes his desperate networking activity and I go back to teaching, my schedule now stepped up to an exhausting twelve hours a day, punctuated with brief intervals in which to rush home to buoy up Marcello's sinking spirits.

I am thinking back over our long marriage one morning on the way to work, reminding myself that Marcello and I pledged to stay together for better or for worse—life having taken a definite turn for the better since Marcello hurtled into it—but inevitably there must be a worse, which is now. My route takes me by the austere Romanesque Basilica of Sant'Ambrogio, and I overhear a funeral congregation chanting an invocation to the saints: *"Intercedete per noi."* Ah, the saints. Thanks to my Protestant upbringing I have had little to do with this host of genial dead personalities hovering somewhere between Man and the Maker, obligingly ready to intervene in mortal affairs. Yet in Italy they are omnipresent. If I so much as cough in the presence of my Milanese friend, she urges me to invoke Saint Blaise, patron saint of the throat. When conjunctivitis strikes, my mother-in-law instructs Saint Lucy to do her bit. A student told me that when an exam is near, he prays to a curious saint—Joseph of Copertino, noted for his powers of levitation, but also endowed with a remarkable facility for impressing examiners. I even discovered from an inscription in a Milanese church that Saint Sebastian, always depicted pierced by arrows, is the patron for the most unpopular category of all, the *vigili*—the Milanese traffic police. With Pope Giovanni Paolo II breaking papal records for beatification, the saints are multiplying rapidly. Somehow Saint Isodore of Seville has been elected as patron saint of the computer, and I read in the *Corriere Della Sera* that a Catholic website is advertising for nominations for a patron saint of the Internet.

It occurs to me that a leg up from a saint would certainly come in handy in this difficult period of our lives. But which one? How does one go about choosing a saintly protector? As a resident of Milan, perhaps I should look no farther than the crypt of the adjacent Basilica, wherein lies the mummified body of the city's patron Saint Ambrose, first Bishop of Milan. Many a time have I accompanied visitors to gaze on his desiccated relics displayed in the crypt, but somehow I am not inspired to address my problems to him. Is there a patron saint for the families of unemployed businessmen? If such a paragon existed, he would surely be oversubscribed, and, if at all effective, would create absolute havoc on the Saint Market Economy.

With Marcello still out of work, the dream of Santa Maria has settled to the very basement of our minds when Fabrizio, an old friend, turns up at our flat and offers Marcello a full-time executive job. This is wonderful news, and it comes as an enormous boost, not just to our finances but, more importantly, to Marcello's battered self-esteem. Although it does not bring us back to the income level of yesteryear, it gives us hope. Did the saints have anything to do with this, I wonder? Have they been intervening for us behind the scenes?

With the new job security and what Marcello calls "our little cushion" nicely plumped by a bullish stock market, the subject of Santa Maria comes up once more. Perhaps we can have our dream home in Sicily after all. Once the idea takes hold, it gathers sudden urgency. We call Silvio to sound him out. Used to our changes of mind he is naturally wary, especially when we tell him we want the ground floor finished and ready to move into by the following summer. This time, we assure him, there will be no going back. Like all architects, Silvio is loath to commit to a time scale, but in the end he concedes.

"A year should be enough time, provided we don't meet with too many *imprevisti*."

"Imprevisti—unexpected? This was a word we came across many a time during the unfolding of the Cascina Fiasco, and it was a word that always seemed to bring trouble. Still, we trust Silvio not to exploit the imprevisti. No one could be better qualified to oversee the work and besides, he is Family. Thanks to his position in the Soprintendenza, the regional Cultural and Environmental Office, he is well acquainted with local building contractors. He will get the estimates and decide on the best offer.

Meanwhile, Marcello and I are beside ourselves with excitement. We are well aware that the golden rule in building projects is to be there on the spot, to know exactly what is going on at all times, as we would have done with the cascina had it not become a fiasco. But Mazara is a long way from Milan, over six hundred miles in fact. Since supervising the project ourselves is out of the question, we will have to put our faith in Silvio. We decide to go down to Sicily at the first opportunity, which will be at the beginning of November.

The first of November, All Saints Day is usually a public holiday. This year it falls on a Sunday, which means that the Italian populace is cheated out of the extra day or two that usually make a *ponte*—a "bridge," or long weekend. However we will extend out stay to give us time to discuss the project and leave us free at the weekend.

At lunch on All Saints Day the topic of conversation is Padre Pio. The sickly priest of San Giovanni Rotonda in Puglia received the stigmata in 1918, and since his death in 1968 many miracles have been attributed to him. His recent canonization has gripped the nation in a frenzy of Padre Piomania. Unwittingly, he has become the champion of the *abusivi*—the illegal builders, the most notorious of whom have built homes in the protected zone near the Vale of Temples at Agrigento. Television news bulletins have been showing bulldozers poised to attack—not just villas but an abusiva church, dedicated to—who else but Padre Pio? The protestors refused to budge from the threatened houses and wave the saint's banner at the bulldozers poised to strike.

"Padre Pio, intercede for us," they wailed aloud to the saint and the television cameras.

Nanette's mother, Anna, has her own ideas on the showdown taking place at the Vale of Temples.

"Padre Pio's got his own chapel up there. He's gotta be on their side."

Silvio's thirteen-year-old son, Massimo, says he's fed up with Padre Pio. He has just taken a Greek exam that the whole class had been dreading. Previously his schoolmates had brought in good luck charms, amulets; or pictures of pop stars to help them get them through such trials. This time nearly every candidate had a picture or a statue of Padre Pio on his desk.

"Did you pass, Massimo, even without the help of Padre Pio?"

"Of course I did, and anyway, if I was going to pray to a saint, it wouldn't be Padre Pio. I'd go for Sant'Uberto." Silvio nods, clearly approving his son's choice of the patron saint of hunting. Eleven-year-old Stefi pipes up, "Better San Francesco, who loved the little birds."

I also prefer the green Saint Francis to Saint Hubert, the stag hunter. I'm just about to get going on the subject when Silvio reminds us that you have to be careful if you rely on Saint Francis to perform you a miracle. Have we heard the story about the fisherman who made a penitential vow to Saint Francis and was then drowned at sea? All because he got his Francescos muddled up and prayed to Francis of Assisi instead of Francis di Paola, who walked on water.

Before an argument over saints can get under way, Nanette says her mother has something to tell us about Padre Pio. Anna, a Sicilian American whose mother tongue is English, demurs at first, but Nanette urges her on.

"Go on Mother, tell them what you saw."

"I saw him, I saw Padre Pio. I really did!"

"Go on Mother. Tell Caroline and Marcello, tell it all." Everyone else has already heard the story by now, it turns out, although Anna didn't tell her daughters until months after the event. Nanette was hurt by this, "Gee, Mother, you could've told me, your own daughter!"

"Yeah, well, I didn't want you to think you had a crazy old Momma." Anna has almost lost her sight. She tells us she had been sitting on her bed, thinking about a book she had once read about the saint, when she decided to speak to him.

"Padre Pio," I said, "I'm so glad I read that book about you when my eyes were still good. I'll never read again, Padre. I just wish I could've met you myself. And then there he was, Padre Pio himself, just hovering in the doorway."

"What was he wearing? How did you know it was him?"

"You can't mistake Padre Pio. I just knew; he was wearing brown, a long brown robe."

"Did you see the stigmata?"

"The what?"

"The nail marks on his hands and feet?"

"No." She pauses on this one. "I don't remember anything like that. He was all in brown."

Nanette butts in, "How could you see him, Mother, when you don't see good any more?"

"I know I don't see too good, but I sure saw Padre Pio that time."

"So what did he do? Did he speak?"

"No, I told you he said nothing. He just hovered awhile and vanished. Then I was so excited I rushed to tell Joe, in the other room, and do you know what Joe said? He said, 'Anna, Anna, you just missed Padre Pio!' " So Joe had seen him too, at the same time but in a different room.

No one challenges Anna's story, and she sits back in her chair, proud of her vision, but puzzled as to why she and Joe should have been singled out for this privilege. Over coffee I ask her whether she thinks that if she had still been living in San Diego she would have had this vision.

"Oh, I guess not, Caroline. Padre Pio's a sick man. He'd never get to travel so far."

WE HAVE COME TO SICILY TO LAUNCH A BUILDING PROJECT BUT incidentally we find ourselves among *I Morti*, the Dead, who follow close on the heels of *I Santi*. The first time I was asked "What are you doing for I Morti?" I was somewhat taken back. What was I doing for the dead? Nothing, I was about to say—I mean there's not much one can do for them, is there? Before I came out with some such reply I realized that I was actually being asked how I was going to spend the Festa dei Morti—the All Souls' holiday, often simply referred to as I

Morti. Although November 2, All Souls' Day, is no longer a national holiday, it has lost none of its importance for families all over the country who remember their Loved Ones on this occasion. In Sicily it is one of the most important celebrations of the year, second only to Easter. With scant respect for the official calendar, the people of Mazara close shops and offices anyway in order to spend the best part of the day at the cemetery.

The prospect of being in Sicily for I Morti for the first time in my life sets me thinking about the Sicilian obsession with death, which is so well known that it can almost be considered a cliché. Just one year before he was brutally murdered by the Mafia, Giovanni Falcone wrote the following:

> The culture of death does not belong solely to the Mafia; all of Sicily is impregnated with it. Here the day of the dead is a huge celebration. Solitude, pessimism and death are the themes of our literature, from Pirandello to Sciascia. It is almost as though we are a people who have lived too long and suddenly feel tired, weary, emptied, like Tomasi di Lampedusa's Don Fabrizio.

Indeed, *The Leopard* opens with words from the recital of the Rosary "Nunc et in ora mortis nostrae—now and in the hour of our death." Frequently during his aristocratic reflections, the Prince admits he is courting death. From his dressing room he muses on the theme.

> From the Mother Church next door rang a lugubrious funeral knell. Someone had died at Donna Fugata; some tired body unable to withstand the deep gloom of Sicilian summer had lacked the stamina to await the rains. "Lucky person," thought the Prince, as he rubbed lotion on his whiskers. "Lucky person with no worries about daughters, dowries and political careers." This ephemeral identification with an unknown corpse was enough to calm him. "While there's death there's hope," he thought.

While most of us are perfectly happy to watch variations on the theme of death played out on the screen, but loath to have any firsthand contact, in Sicily the medieval familiarity persists. Plague, famine, and war were scourges common to most of Italy and much of Europe in the Middle Ages, but Sicily got all of these in greater doses, not to mention earthquakes, volcanic eruptions, and, the final scourge, the Mafia. It was in Sicily that I came across death face-to-face for the first time in my life. Having grown up in a culture that shies away from death, I had never even attended a funeral before Goffredo died, just three years after Marcello and I were married. We were staying in Rome at the time, about to take a flight to Brazil, when we received a call from Lallo.

"Something terrible has happened. Come at once."

We arrived at Santa Maria to find the great wooden doors of the courtyard draped with purple swags and the villa swarming with people. Goffredo was laid out on the matrimonial bed, dressed in a dark suit, his face waxen and drawn from the final illness. It was a sight I was totally unprepared to deal with. He had died in the early hours of the morning, and according to Sicilian tradition, the family was to keep the wake until he left the house for the funeral the following day. All that day, Maria, Pilli, Marcello, and Silvio sat by the bed on chairs placed against the wall, as visitors arrived to offer their condolences, kiss Goffredo on the cheek, and weep. Hour after hour the room was filled with the low murmur of voices and muffled sobbing.

In Sicily death is a public event, and I soon realized that the obligatory rituals left no place for privacy. Distraught as we were, the Sicilian mourning code did not allow us to go away and grieve alone. We had a public role in the community, which meant we had to show our grief to the world. This rule was not rigidly applied to me. As *famiglia acquisita*—acquired family—and with a baby to look after, I was allowed to stray and wander around the house, but not into the garden (a restriction that I found inexplicable at the time, although I later realized it was because the garden was visible from the road, and it would not do for a member of the family to be seen to be distracted from the mourning rites). The strain of taking part in the scenes of grief was sometimes too much for me, and every now and then I would sneak off with Melissa to join Antonietta and the group of former retainers in the kitchen. Cooks, nannies, and the wives of ex-caretakers and gardeners formed a sisterhood of elderly widows dressed in black. Entranced with the bonny baby holding court on the kitchen table, they swore she was the spitting image of her father, and I had to agree there was a similarity between the curly-headed Melissa and the photograph of the two-year-old Marcello, dressed as a girl with ringlets and ribbons.

Although Melissa provided a welcome diversion, I simply could not come to terms with the terrible sense of loss at Goffredo's death. I had warmed to him as soon as we met, and I believe the *simpatia* was mutual. My membership of the race of Albion automatically endeared me to him, and besides, he liked my low-key ways, an antidote to the rest of the family's exuberance perhaps. In me he found an ally in the futile battle against the *prepotenza* of his wife and his eldest son, and I could always be counted on to laugh at the jokes that the rest of the family had heard many times before.

And so I became acquainted with the elaborate rituals that accompany the Sicilian way of death. According to Sicilian custom, the dead body cannot be left alone even for a second during the wake, so at night close members of the family took turns to *vigghiari u mortu*—to watch over the dead. Despite the chill of the night air, the bedroom window was left wide open to allow Goffredo's spirit to depart. We conducted the vigil in pairs, and it was my turn to keep the wake in the small hours of the night together with Zia Antonietta, the widow of Goffredo's brother, Valentino. This was the first time I had met this traditional Sicilian matriarch, and in these sad circumstances we struck up a close friendship. In those days with the help of her son, Luigi, she managed a large farm of vineyards and olive groves. We talked into the small hours as she recounted anecdotes of the Manzo family and of her work on the family farm at Rocazzo. When it was time to call Marcello and Pilli to take over the vigil, we decided to let them sleep while we carried on talking at the foot of Goffredo's bed until the sun rose.

With the whole family in mourning, I alone had no idea how to behave. There were rules of which I was ignorant, but which everyone else instinctively adhered to. I knew that Maria would wear black, but cosmopolitan and cultured as she was, I never imagined she would conform to the stereotype of the Sicilian widow and wear it for the rest of her life. Nor did I immediately realize that I too would have to conform to the rules of mourning. On Franzina's gentle insistence, I put away my unsuitable summer dresses and donned a dark frock of Maria's. Perplexed at the fact that even Melissa had to wear a black armband, I assumed the dress code would be relaxed after the funeral. Surely they didn't expect me to wear a black dress when I took Melissa to the beach. Maria Antonietta took me aside and explained. There was no question of taking Melissa to the beach. None of us were to leave the house for nine days, the official mourning period. What about the shopping? How would we feed ourselves? All would be taken care of. Our food would be brought by the Zie and Cugine. There would be no need for us to scendere to Mazara, and no need to cook. We, the close family, would stay at Santa Maria to devote ourselves to the memory of Goffredo and receive visitors who came to offer their condolences. Why nine days, I asked, and not seven or ten? My sister-in-law didn't seem to know.

After twenty years of immersion in the culture, it seems to me that Sicilians are not so much obsessed with death but, unlike the rest of us, are perfectly at home with it and treat it as the natural phenomenon that it is. Like the ancient Romans, they acknowledge that the dead are gone from us physically but their memory is ever with us. Conversation at the dining table is often about dead friends and relatives, as though their actions and opinions were still of vital importance to all concerned. If life does not actually continue after death, the distinction between the two states is certainly blurred. On the subject of the rituals of dying, Leonardo Sciascia recalled attending the deathbeds of relatives when he was a child in the 1920s. The family would gather round the dying, not only to keep them company in their last hours, but also to make raccomandazioni, asking them to carry messages and bulletins of family news to the dead souls they would meet in purgatory. On one occasion there were so many raccomandazioni of this type that the dying man in exasperation begged everyone to write the messages down on pieces of paper that he could take with him into the afterlife.

Sicilian children are used to the constant presence of grandparents, and the link with them does not end with death. When a relative dies, there is no attempt to conceal the fact from children, who are usually allowed to see the body and attend the funeral. All Goffredo's grandchildren came to his funeral, including Melissa, then eleven months old. As the baby of the Sicilian family before Melissa's appearance on the scene, Emiliano was his grandfather's pet. Who knows what was going through the mind of the four-year-old who had attended the wake and was taken to the funeral? Apparently serene, he busied himself picking daisies from the garden for his *nonno* and brought them to the cemetery. Clutching his flowers, Emiliano watched as Goffredo's coffin was carried into the Vaccara family vault. Instead of placing the flowers next to it, he followed the crowd outside then suddenly threw them up in the air. When they fluttered to the ground, he burst into tears. "If Nonno is up there in heaven, why doesn't he catch my flowers?"

Sicilian children have always been familiar with death. Until recently the most exciting day in their lives was not Christmas or birthdays but the *Festa dei Morti*. There was nothing spooky or frightening about this festival, which marked a special closeness between the young and the old. On the eve of November 2, children awaited the arrival of their dead relatives. The spirits would rise from their graves in the cemetery and make their way in the dark to the town nearby. Arriving at the town, they would fill the streets and alleyways, seeking out their old homes where the children of the house lay trembling with excitement in their beds. Only when the little ones were asleep, however, would the spirits climb up the walls of the house. On the balconies and at the windows they would leave their gifts, the humble offerings of the contadini—almonds, dried figs, prickly pear conserve, little jars of colored sugar. The most lavish presents were the *pupi di zucchero*—puppet-like dolls made of crystallized sugar, which still appear in the local pastry shops for I Morti. Some take the form of girlie ballet dancers in pink or peasant girls in costume, like the two-dimensional cardboard dolls we dressed with paper cutout clothes back in the 1950s. But there are also dólls with macho appeal—the paladins, Orlando and Rinaldo, and Saracen warriors lifted directly from the Sicilian puppet shows. The lineup of sugar celebrities include Puss in Boots, Snow White and her dwarfs, not to mention the bearded Garibaldi, with his red shirt and poncho, a motley assortment of characters from Sicily's past and present culture. In addition to the *pupi* are the marzipan fruits traditionally offered in honor of I Morti. Ever since the nuns of the Martorana convent in Palermo inveigled their guests into plucking marzipan oranges and lemons from the trees of the convent, Sicilian pastry cooks have delighted in deceiving the eye with almond paste concoctions. Dusky peaches, shiny purple plums, prickly pears, and bananas complete with bruises are all rendered in virtual marzipan version. Along with the sugar puppets, these were the presents children would look forward to on the Festa dei Morti. But Sicily being Sicily, the pastry cooks could get carried away. By the eighteenth century they were bored with virtual fruit and were shaping the almond paste into more macabre forms. One archbishop of Palermo had to issue a ban on the manufacture of *ossa di morti*—marzipan fashioned in the shape of crossbones and *cruzziteddi*—dried chestnuts decorated as skulls. Nowadays the Pasticceria Lamia, where we go to buy cassata and cannoli for special occasions, stops short of the macabre but displays some amazing examples of marzipan confections of shrimps, octopuses, and crabs. The prize piece is a giant lobster, fashioned in lifelike detail down to the last claw, garnished with wedges of Sicilian lemons, complete with shriveled pips.

Marcello and his brothers belonged to the last generation of children to be visited by I Morti. In their case the gifts were not left at the window but hidden around the house, so that the anticipation of gifts was heightened by the thrill of the treasure hunt. But the gifts were still the traditional sugar puppets and marzipan fruits that his parents and grandparents had received as children. When Clem first heard Marcello recounting these highlights of his childhood, she was amazed.

"Sugar dolls and marzipan! Papà, didn't you ever get anything more exciting than that? I mean, didn't you get any proper presents?"

"No, we thought ourselves lucky to be getting any presents at all."

Marcello and his brothers never made the connection between the value of the treasures I Morti had hidden and what his parents could afford. Nor did they find it scary that the gifts were ghost borne. Certainly it seems more logical to receive presents from kith and kin, even if they happen to be dead, than from a bearded old stranger. Besides, a reindeer-drawn sleigh is a bizarre mode of transport in the Sicilian context. For those of us who cherish history and tradition, within one generation globalization has brought Father Christmas to Sicily as major bearer of gifts so that Sicilian children have lost a precious local tradition and adopted the culture of children the world over from Sidney to Sarasota. More recently, Halloween has also arrived, completely upstaging what was left of I Morti.

ALTHOUGH THE CUSTOM OF THE DEAD BRINGING PRESENTS TO the children is virtually extinct, the feast day itself still remains one of the most important in a culture that reveres their dead ancestors even more than their saints. In Sicilian tradition we will be paying respect to the recently dead close members of the family. Our first visit today will be to Goffredo. Until some years ago he had been interred in the Vaccara chapel. By tradition he should have gone straight to the Manzo vault in Trapani when he died, but Maria always maintained that she herself would be buried in Mazara with her own family, and where she went, Goffredo would go too. My mother-in-law's flouting of the rules led to a family dispute. The Vaccara chapel had been designated for

Luigi's sons and their wives, and Goffredo could not take the place reserved for a Vaccara. While Maria's brothers and sisters-in-law were still alive, the issue was shelved, but as they began to die off one by one, it became urgent. No one had the courage to tell Maria that Goffredo would have to be removed from the Mazara chapel and transferred to Trapani, and, moreover, that she would eventually have to join him there. In the end Goffredo was moved to Trapani, but only when Maria was in the advanced stages of Alzheimer's.

Italian cemeteries are like walled cities, with tree-lined roads and rows of family chapels built to house several generations. The facade of each chapel makes a public statement, the quality of the architecture and especially the materials, marble, brick, or cement revealing the wealth and status of the family. Inside, the walls are lined with tombs, each with a touching photograph to remind later generations of what their *nonni* and *bisnonni* actually looked like. These mausoleums reflect the full range of hyperboles in Sicilian architecture, from baroque pyramids of angels to Greek temples. The scene would be lugubrious were it not for the buzz of the crowd swarming around us. The paths are seething with people, whole families on an annual outing, teenagers arm in arm, couples with strollers, and children skipping behind. Groups of people standing around to chat block the pathways so that we have to step off into the dirt to get by. The dress style is surprisingly uniform with elderly widows and teenage punks alike dressed in black from top to toe. Occasionally a bare midriff or a flash of piercing proclaims the spirit of youth. Stray dogs bask beside tombstones. If it weren't for the graves, we could be in a city park on a Sunday morning.

Having followed Silvio's directions assiduously, we still manage to get lost. We ask for directions to the Manzo chapel, but nobody seems to know it. On the fourth or fifth enquiry we strike it lucky. There it is, with the inscription over the wrought-iron gate built in 1926 by Goffredo's eldest brother, Francesco Manzo, notary of Trapani. Like many of its neighbors, this chapel is decorated with the stylized eagle wings that I associate with Milan Central Station, also built during the Mussolini era. The tombs are stacked to the left and right of the entrance, and a crucifix hangs on the facing wall. Below the cross is a console with framed pictures of the family, as you would expect to find in any middle-class sitting room. The Manzo family patriarch, Luigi, Goffredo's father, had eight children: five sons and three daughters. The daughters are not entombed here but have gone to the chapels of their respective husbands; this chapel contains five Manzo brothers and their wives. Most of the pictures are youthful images, the faces that the family wants their loved ones to be remembered by. The inscriptions are full of clichés—so many loving husbands and wives, doting fathers and mothers, dutiful sons and daughters. Could they really all have been such paragons of virtue?

The flowers that tradition prescribes for the cemetery are chrysanthemums, which I happen to loathe, so we have brought roses and daisies instead, roses for Massimo and daisies for Goffredo, who shared Maria's passion for them. We arrive to find the chapel open. The wife of one of the Manzo cousins is busy filling vases with water and throwing out the old flowers, while another is sweeping out the dead leaves in preparation for the arrival of visitors. The women are delighted to see us and set down their tools immediately for a chat. Soon afterward more Manzo cousins arrive, sons and daughters of Zia Antonietta, who has recently died and has taken her place in the chapel. Sadly I reflect on how all those years ago she and I stayed up all night keeping the wake for Goffredo. Now she lies in the tomb opposite him. The last time we met, a family feud had been raging, but for Zia Antonietta's sake her children seem to have made peace, and the hatchet, if not exactly buried, is at least semi-interred. However, the out-of-favor brother does not engage in conversation with his siblings. After the preliminary salutation, he pulls out a do-it-yourself kit and starts drilling into marble. Conversation is drowned by the roar of the Black and Decker as he drills on. His wife explains that he wants to hang up a photograph of his mother to go over her tomb. A hole has to be drilled, but perhaps today wasn't the best day to do it, she shouts over the din.

We move outside into the sunshine. Somehow this has turned out to be a social occasion for the living rather than a quiet contemplation of the dead. I had hoped for a few quiet moments at the tombs of the three Manzos who had been dear to me, Goffredo, his favorite brother, Massimo, and Zia Antonietta, but awed by the ear-splitting din of the drill, I find this is impossible.

Outside, the group expands as more Manzos gather to pay their respects to their loved ones. Silvio and Nanette arrive with Massimo and Stefi, and we now have a reunion between the Trapani and Mazara branches of the Manzo family. At first we all stand around solemnly as the occasion demands, heads bowed, eyes down, reminiscing about ancestors, but within minutes the conversation lightens up considerably, with two Manzo cousins holding forth on this year's olive harvest, which has just begun. Both have their own olive groves, planted with the Nocellara di Belice variety (considered the very best in Sicily for extra virgin oil). Silvio tells them he has just planted olive trees on his land at Santa Maria, and Marcello and I announce our ambition to join the ranks of oil producers, once the house project has been completed. Luigi Manzo, Zia Antonietta's eldest son, agrees. The terra of Santa Maria is excellent for olives, but if we want his advice, we should eliminate the rows of cypresses that border our land. Cypresses and olives don't go together, he insists. We have no intention of chopping down trees planted by Maria and her father before her (in any case the olive trees can be planted at a suitable distance from the cypresses), but we nod, pretending we will take his advice. Silvio and Marcello place orders for Luigi's oil from his farm at Rocazzo, and a debate ensues, if you can call it a debate, when there is unanimous agreement from the start, over the merits of the local extra virgin. No doubt about it: Sicilian, that is Sicilian from the province of Trapani, is the very best, far better than anything you can get from Tuscany or Puglia. There has been a lot of publicity recently about oil from Liguria. What does Luigi think? He chortles and coughs. In Liguria, they use *nets!* Using nets for the harvest means that the fruit gets bruised before it goes to the mill. Sicilian olives are handpicked. Other regions of

Italy may get more publicity for their olive oil, but we in Sicily *know* ours is the best. Both Manzo cousins are quite emphatic on that point, although I very much doubt whether either of them has ever bothered to sample oil from anywhere else.

Eventually the party breaks up. If we had more time we would take up the cousins' invitation to lunch, but we want to be back in Mazara by the afternoon to pay a visit to the Vaccara vault, especially to Omar. He was the second child of Pilli and Maria Antonietta, and his death from cancer at the age of thirteen was one of those terrible events that shake one's belief in God. To lose a child is the worst tragedy that can befall any parent, but to lose a child as enchanting as Omar was devastating. He was three years old when I first met him, and he totally captivated me. I was a newcomer and a foreigner—I spoke funny—he said, but he adopted me anyway, introducing me to his younger brother. "This is Emiliano, he cries a lot, but he's okay." Coming back from the beach it was Omar who hosed down our feet in the courtyard, making sure we were sand-free to enter the house. When we climbed the flight of stone steps to the upper floor, he would go in front, turning every now and then to remind me in lisping solicitation "*Stai attenta!*" Years later, ever mindful of younger children, he came to the rescue when Melissa was stung by a Portuguese man-of-war, obligingly rushing over to pee on her leg, ammonia from urine being the best relief for jellyfish stings, as every Sicilian child knows. Omar was one of those children who bring extraordinary pleasure into the lives of others. He was exceptionally bonny and beautiful; he must have cried now and then, as all children do, but I never saw him lament, even in the last year of his life, when he was often in severe pain. During his final illness his parents gave him a dog, a mutt he called Max, and I think his longing to get well, to be able to get out of bed and run and play with his new love, buoyed up his spirits, although it was clear to everyone else that the end was near. The last treatment he underwent was at a specialist pediatric hospital in Pavia, not far from Milan, where we were living at the time, and Pilli and Maria Antonietta stayed with us for much of the time in those last weeks of their son's life. Looking back, I am amazed at the fortitude they displayed when they knew that his death was so near. He died in the hospital, and when Pilli called that fatal night with the tragic news, we knew that one whom the gods had loved, perhaps to excess, had been taken from us, and for none of us would life ever be the same again.

The *Campo Santo*—walled cemetery, literally "holy field"—of Mazara is a more cheerful place than its counterpart in Trapani. As at Trapani the outer walls are lined with flower stalls bedecked with the ubiquitous chrysanthemums, but inside there are fewer family chapels and more graves, so, despite the cypress trees, it has something of the atmosphere of an English churchyard, with lawns and flowerbeds carefully tended for the occasion. Omar lies in one of the grand chapels lining the walls of the cemetery, his marble tomb inscribed with the dates August 1973 to April 1986. A photograph of a sunny eleven-year-old in a Scout's uniform unleashes a host of memories and not a few tears. Fighting them back, we lay a bunch of daisies at the door of the chapel and leave.

THE GROUND FLOOR, 1970S

NINE

Tales of the Unexpected

PER SAN MARTINO, OGNI MOSTO È VINO.

Early November in Sicily is known as Saint Martin's summer, for the warm weather generally persists until his feast day, November 11, when the "must"—the grape concentrate—of the previous harvest is transformed into wine and ready for the tasting. The days leading up to this festival are, as Tomasi di Lampedusa puts it, "The real season of pleasure in Sicily: weather luminous and blue, oasis of mildness in the harsh progression of the seasons, inveigling and leading on the senses with its sweetness, luring to secret nudities by its warmth."

I don't know about secret nudities, but there is certainly plenty of warmth, and we have arrived from Milan hopelessly overdressed for a Sicilian autumn. As I stripped off layers of wool and tweed the day we arrived, I scolded my husband for not having warned me. Surely he should have known how hot it could be in Sicily at this time of year. But no, apparently he has lived away from the island too long to remember such things. I note with envy that the locals are still in shorts and T-shirts, and as we drive past the beach of San Vito, we spot bathers in the sea. Still, there is no time for idle pursuits such as swimming or sunbathing; we have much work to do. First we must get our heads around Silvio's figures.

When we were engaged on our ill-fated cascina project, the architect presented us with a neatly typed list of all the potential costs along with floor plans and cross sections. It was easy to see the difference between the architect's estimates and those that eventually came in from the contractors. With Silvio there are no diagrams and no computer involved—simply strings of numbers in illegible handwritten scrawl, which he has already faxed to us in Milan. As neither of us can work out the sums in what he calls the *computo metrico*, we must get him to go through it and explain it to us face-to-face.

But even face-to-face, huddled around the papers scattered over the dining room table with a glass of the best Marsala for support, we are none the wiser. Befuddled by the complexities of the computo metrico, Marcello tends to opt out, leaving me to thrash out the details with Silvio. I am no more equipped than my husband to ask intelligent questions, much less to raise objections. I do notice, however, that what we had assumed to be a perfectly good floor in the Flowers' future bedroom is destined for *rimozione*.

"Silvio, how could this be?"

"You need a bathroom in this area, *è giusto?*"

"Yes . . . ?"

"And bathrooms need plumbing, *è giusto?*"

"Yes . . . ?"

"And plumbing goes under the floors, *è giusto?*" I see what he's getting at. It is indeed quite *giusto.* "So the pipes have to go under the bedroom floors, you mean?"

"È logico!" He has finally got through to me. Silvio explains that if we want new bathrooms, both the disused well pump and Maria's dog kitchen will have to be demolished. There does seem to be a lot of demolition for what was supposed to be a conservative project. But the arithmetic involved in calculating *demolizione, dimissione,* and *rimozione*

in cubic meters is as complicated as it is boring. In the end I decide to leave it to Silvio, reminding him that *economia* is the buzz-word. But even with the purchase of new materials kept to a minimum, the project still looks dauntingly expensive. *Preventivo*—estimate is another word that pops up frequently, more often on Marcello's lips than on Silvio's.

"Silvio, where is all the money going?"

"*Beh, il problema è l'umidità.*" Humidity? What a joke! On the one hand Sicilians are always lamenting *la siccità*—drought—but now it seems the problem is Damp.

"Why do you think the walls are crumbling away? Santa Maria has no foundations, and there is a waterlogged clay table underneath, so the *umidità* just seeps up through the walls."

"What can we do about it?"

"Well you could put in a false wall, or you could rebuild the outer walls and inject them with special antihumidity agents." He goes through various options for battling the damp, all extremely expensive and none guaranteed to work.

"To tell you the truth, if I were you, I wouldn't spend money on any of that. But you are going to have to scrape down most of the walls and cover them with at least two layers of plaster. We may find that the ceilings need doing too. I'm afraid the cost will be *non indifferente.*"

Silvio's *non indifferente* turns out to mean nothing less than staggering. He pulls out the computo metrico and starts doing more sums in the margin.

"Now if we leave the ceilings—and that's being optimistic—and a slice of the walls at the top just as they are and just touch up any trouble spots, I think we could make a saving. *È giusto?*"

"Right, but even so, we are still talking thousands of square meters, *è giusto?*" I've only been back in Sicily for a couple of days, but I'm already picking up the speech pattern.

"*È logico.* But don't forget, the cost of labor here in Sicily is less than half what you would expect to pay up north."

Silvio's choice of building contractor has fallen on Signor Cassano, who has worked for the Soprintendenza at Selinunte and proved to be both reliable and reasonable with his prices. After lunch he comes to Santa Maria to discuss the project over coffee. I take to him instantly. He is very dark, with one of those solemn Byzantine faces that stare down from the church mosaics the Normans commissioned from Greek artists. With his oval face and slanted eyes, he could have been the life model for the Pantocrator at Cefalù or Monreale. It is not difficult to imagine him with a long flowing beard. He has a slight speech impediment, which makes us pay close attention to what he is saying. He is keen to undertake the Santa Maria project, whether driven by an enthusiasm for old buildings, or because it was his cousin who did the botched plastering job in the eighties and he feels he has a debt of honor to pay on this account, is not entirely clear. He says he will make it up to us by restoring the walls "*a regola d'arte*—to a fine art." He is confident he can eliminate the damp by using materials that will let the tufa bricks breathe.

My limited vocabulary in Italian has let me down in the past; this time it threatens to jeopardize a budding relationship with a Byzantine builder.

"And Signor Cassano, you will give us plenty of notice before you start work on the bathrooms. So I have plenty of time to order the *pipistrelli*."

"*Piastrelle*, you fool," hisses Marcello. "Not *pipistrelli*. You've just told him you want bats in the bathroom." As if one malapropism were not enough for this crucial encounter with the man we are entrusting with our dream project, I put my foot in it once more, this time on the subject of the Damp.

"Signor Cassano. I know we can't do anything about the foundations *under* the villa, but Silvio suggested a *vespasiano* on the garden side. Would it be very complicated? What do you think?" The sombre Signor Cassano can hardly keep a straight face. Silvio smiles indulgently and Marcello bursts out laughing.

"Now Caro*line,* do tell us. Just why would you want to put a *urinal* alongside the villa? To banish the damp?" The word I was aiming at was not *vespasiano* but *vespaio*—honeycomb drain or damp course. Eventually I learn from Silvio what I had managed to miss in the course of a classical education—that the emperor Vespasian was famous for introducing peeing facilities all over Rome, for which he has the honor of bestowing his name on the modern public urinal.

Like Silvio, Signor Cassano is full of talk of *imprevisti*, but is not so keen on *preventivi*. When it comes to his estimate, he declines to go into the nitty-gritty details. He and Silvio will do as they do at the Soprintendenza—establish a fixed price for each job and then work out the cost step by step as the work proceeds. We realize it is no use trying to sail against the wind by insisting on an overall estimate in advance. The job rates are reasonable, so we will cross our fingers and pray we will be delivered from *imprevisti*.

What will we do for water? We have assumed we can be linked to the municipal water system, but apparently this is not the case. Silvio tells us it will be prohibitively expensive. Santa Maria is only three miles from Mazara, but it is isolated, and the town's aqueducts stop at least a half a mile away. If we want a connection, we will have to pay for the connecting pipeline ourselves, in addition to the connection charge. Even then the continuous supply of water would not be guaranteed. Sicily is famous for *la siccità*, which deprives hundreds of towns of water in the summer months for weeks on end. The problem is not lack of supply but inefficient distribution. Protests to the local authorities have no effect, and other than an appeal to the local patron saint, what else can be done? Rather than depend on capricious saints or the comune of Mazara for our water supply, we will have to do what everyone else in the countryside does and dig our own well.

MONTHS LATER, AT THE TAIL END OF WINTER, I AM LYING IN BED in Milan convalescing from the latest in a series of hip operations when Silvio rings with news, good and bad. He gives me the good news first.

"Eureka! We've found water."

"Just like that?"

"Yes, the water diviner was here this morning. He identified the spot immediately, and Cassano's men hit water at twenty-five meters. They're already digging the well."

"And the other news, Silvio?"

"There's an old septic tank under your kitchen leading into an underground tunnel. We had no idea it was there. But anyway we'll have to block it up and extend your sewerage system out into the garden. Your tank will have to go into the orchard. That means extra costs. I told you to expect a few *imprevisti*.

"Okay Silvio, but what about the tunnel? How long has it been there?" In my mind's eye I already see buried treasure and jewel-encrusted chalices hidden from raiding Saracens.

"Silvio, this is not bad news. A secret tunnel! This is exciting!" But Silvio is not excited.

"There's no way anyone can explore this tunnel for one simple reason."

"Which is?"

"It's full of . . ." He's too fastidious to say the word.

"Merda?"

"Esattamente. It's blocked solid and probably has been for centuries. All we can do is seal it."

So much for our secret tunnel. We try to convince ourselves that the discovery of water so easily is of more practical value than any hypothetical treasure, though the question remains debatable. No good musing at length about it. The main thing is that now we have water, the work can go ahead, which it does with amazing speed. Every night we call Silvio from Milan to get the blow-by-blow account. At first it is mainly the demolizione and rimozione. Blasting out the well pump and knocking down the surrounding walls is a major and expensive operation. Then there are floors to take up to make way for plumbing. I am glad we are not there, but I feel sorry for Silvio and Nanette, who now live on top of a demolition site.

There can be nothing much to see at this stage, but even if there were, I have been put out of action for the time being with an impre-visto of another sort. After umpteen operations to correct congenital dislocation of the hips, I have had both replaced within a year. Compli-cations from the second replacement mean recovery is taking longer than expected, and it will be months till I am rid of the wretched crutches. Every time we speak to Silvio, he asks us when we can come down to see the work in progress.

"And I have a *sorpresa* for you."

"*Una sorpresa o un imprevisto?*"

"No, no, a surprise, a wonderful surprise, and it's nothing to do with the tunnel. Apart from that, you really need to talk to Signor Cassano and the electrician. Various things have come up and it's too complex to describe on the phone." In any case, we don't need persuading to go down to Sicily. We decide to make the trip, crutches and all.

It is March and the rains have come and gone, leaving a light, soft cloak of green over the earth that is bare and parched for the rest of the year. This is Sicily at its most beautiful, just as I saw it first all those years ago. The appearance of *erba* (grass and salad both going by the same name) on this dry Mediterranean island is a good sign; it means the end of the *siccità*. When a Sicilian points to a patch of erba with the words, *che bel verde!*, you are supposed to stop and admire what may look suspi-ciouly like weeds. It is a miracle that the rains have come and graced the earth with green of any sort.

The first thing we do the morning we arrive is go out onto the ter-race to take a look at the orange grove in spring. We were expecting to see it transformed into a sea of green, bobbing with golden orbs and car-peted with wild flowers, and indeed the east side of the garden near the house is gloriously verdant, with its yellow carpet of oxalis, dotted with wild lilies. But on the north side, by the gate that opens into the *parco* outside the villa, unsightly building matter has very much prevailed over nature-loving mind. Part of the admittedly decrepit wall has been knocked down to give access to Cassano's truck, and several citrus trees have been sacrificed to make way for its maneuvers. A cement mixer, piles of gravel, and sacks of cement have claimed the northwest corner formerly occupied by lemon trees. In short, a large chunk of our orange grove has been converted into a builder's yard.

Marcello is visibly upset to see trees planted by his parents and grandparents cut down to pathetic stumps; but the worst is yet to come. The pride and joy of the garden, a vigorous wisteria vine that covered the walls between the house and the gate, its profusion of branches entwined around an abandoned iron gate in a Medusa-like grip. This beautiful plant, untended for many years, was famous for its prolific blossom. Friends and family would come to Santa Maria every spring to take cuttings. Our favorite photograph taken at the time of our wedding features the two of us embracing under its cloud of blossom. One look at the naked wall tells it all. The wisteria is gone. This time I really do wipe away a tear or two. It turns out the plant was in the way and occupied the space destined for the boiler. Silvio sympathises. He was not present when the deed was done nor was he consulted. It is not the first time I have to come up against this Sicilian attitude applied to plants and animals alike, by which the creations of nature are allowed to exist but are promptly eliminated as soon as they start to get in the way. Marcello crouches down and digs around in the rubble till he finds the jagged stump. It is there, rudely snapped by some Sicilian Samson. He rubs it, as though a loving massage will bring it back to life. Yes, it will probably grow again, but it will not regain its former glory, in our lifetime anyway. The English in me wants to challenge the *muratori*, like an irate school prefect, and demand angrily "Now, who is responsible for this?" Fortunately, I have absorbed enough of Sicilian culture to know that confrontation is not advisable.

So when Signor Cassano turns up to show us the sorpresa, we are all smiles. We enter the house from the door into the Flowers' flat-to-be. Straight ahead of us, where there used to be an internal plastered wall with a small arched door, now stands a magnificent Renaissance-style arch, twelve feet high, constructed in bare blocks of tufa. There it is, straddling the entire width of the room, in as perfect condition as the day it was built. What a discovery. It is almost unbelievable. Signor Cassano, all is forgiven.

"How did you discover it?"

"The wall was in bad condition and had to be partially rebuilt and reinforced. During the demolizione we came across the rough tufa, so we just worked at it."

Silvio is also very pleased with the work. He knew that the original design had been based on a series of arches, some of which are still visible on the external walls of the building, but this one is a true surprise. He is sure it is part of the original seventeenth-century construction from the shape of the arch and the quality of the tufa. The *sorpresa* has confounded our original plans for the mini-apartment and sent the architetto scuttling back to the drawing board. His final solution reduces the bedroom area by half, giving the girls a monastic cell to share, but creating an enormous living space. The new layout may not fit the Moorish theme I had in mind, but the magnificent arch is there to stay.

We are looking forward to meeting Signor Crescente, the electrician and plumber (in Sicily, where technicians are more flexible with their job profiles, they are often one and the same thing). He has the reputation of being the very best in the trade, but according to Signor Cassano, he can be difficult to reach. However, he has promised to come to Santa Maria this morning to hand over the preventivo. When he finally appears, Signor Crescente looks less like a plumber than a film star. Perhaps a little too short to make the film-star grade, he has the coloring of a male Snow White: glossy, thick black hair, the whitest of skin, and the scary blue eyes of a Siberian Husky. He uses these to stare us into acceptance of a 200 percent increase in his preventivo.

"I'm sorry, but I don't quite see how it can be three times as much as the original." (This in the most nonconfrontational manner I can manage.) He gives me an exasperated look.

"The first one was just for the disposizione—the laying of wires, pipes, etc.—not the materiale, and it didn't include the central heating or any work in the garden."

"Where is the final estimate?"

"I'm giving it to you now."

Another stare and a shrug. So far he has given us nothing in writing, so there is no way of comparing figures. It is too late to say thanks, I'll think about it, and get another estimate. Everyone is looking at me now—the Architetto and Signor Cassano staring patiently, waiting for the verdict, and Marcello glaring accusingly at me for not having done my homework properly. The blue eyes are staring at me too. Do I discern a threatening look?

"I see. *Va bene.*" What else can I say? The men visibly relax and light up cigarettes, except Marcello who takes out his calculator. This is not all Signor Crescente has to say. Now that the walls have been completely remade, he insists it would be madness to hack into them to place the wires. Signor Cassano, protective of his virgin plaster, totally agrees. The wiring must go under the floor, not round the walls.

"So what's the problem?" I ask.

"The problem is that we'll have to take up all the floors and put in new ones throughout."

I dare not even steal a look at Marcello. I fear for the project or my marriage or both. What they are saying makes perfect sense, but why didn't they tell us before, for heaven's sake? At least we would have been able to plan for it. These two little bombshells will send the budget through the roof. I am prepared to take it philosophically, but Marcello? This project was strictly tailored to leave us a modest sum to supplement our retirement pensions, our *cuscino.* He is not at all happy to see the cushion rudely flattened.

Signor Cassano and his team have in fact done an excellent job on the walls. He has made it a point of honor to remedy the botched job his cousin had done in the eighties. Now we have smooth, dry plastered walls instead of crumbling tufa bricks shedding chunks of plaster. I ask him how he did it.

"Well, Signora, in the end we had to scrape the entire wall surface and the ceilings too. We then used a special porous plaster mix so that any damp coming up through the walls will come through to the surface and not get trapped between the tufa and the plaster."

The entire wall surface? What on earth has this done to the original computo metrico? If the Byzantine has had to scrape and replaster the whole area, all Silvio's complicated calculations for reducing the resurfacing were a complete waste of time. This news about the walls is the third imprevisto in one day. I am beginning to see why Silvio wanted us to come down. Still, the good news is that Signor Cassano is confident about the chances of beating the damp, and although I have feigned dismay at the prospect of all the new flooring, I am secretly delighted that we will be rid of the hideous brown ceramic tiles in the living and dining rooms. What are we going to put there instead? Silvio suggests we continue the terra-cotta theme from the kitchen throughout the house. I am not so sure. Terra-cotta is not typical of Sicilian country houses, and we don't want our Sicilian baglio to look like one of those austere Tuscan country houses, all bare white walls and brown floors. Let's try and keep with the character of the building. What was on the floors before? Silvio says the seventeenth-century floor would have been quarried stone. When his grandfather had it renovated in the twenties, they used *graniglia* tiles, like the ones Silvio has upstairs. Graniglia is an amalgam of marble chips and cement, unconvincingly translated into English in the catalogue as "marble grit." These tiles were at the height of fashion in Italy in the Belle Epoque, especially in Palermo, then a fashionable cosmopolitan city, and they were used extensively in the grand Vaccara house in Mazara. Vito, Silvio's Anglophile friend, the photographer manqué, has been our supplier of all materials so far (he did take over the family business after all), and it looks as though we will have to go back to his shop for another frantic tile-choosing session. But like a deus ex machina, Vito himself turns up at Santa Maria on his scooter, just as we happen to be discussing floors. He has heard we are here for a few days, and he's curious to see the work in progress. Somehow he seems to know that we are going to need more tiles.

"When do you want the place ready by?"

"We were hoping to come here this summer."

"In that case you'd better choose the floor tiles now. Come into the shop this afternoon, and we'll prepare the order." Just like that. Having waited a lifetime for the chance to decorate a home, I now have to make this major decision in a few hours.

In one weekend we have managed to exceed the original budget by almost 50 percent. Marcello, with his mind on the cuscino, is not pleased. In the negotiations that take place between the two of us I have to agree that we will postpone the Moroccan bathroom planned for the girls' apartment. Designed by Silvio with a built in mosaic bath-cum-pool, it was the one concession to luxury in the whole project. One bathroom less will be no real sacrifice, and as Silvio points out, without it the builders will finish sooner. As it is, there is an enormous amount of work to do in three months. The Flowers have already invited friends for July, so we must have our home ready for the summer holidays. Marcello plays tough with Signor Cassano.

"Will we make it by the end of June?"

The answer is reassuring. "Provided I get the materials, when I need them, we'll make it. I have plenty of men to call on. If necessary, I can call them off other sites. We'll have to keep tabs on Signor Crescente though. He has a habit of disappearing."

In spite of his vanishing potential, the Husky look-alike is a model of efficiency. He has a computerized plan of the house and wants my input on where to put power points, light switches, television outlets, and telephone jacks. He advises an emergency light system and convector heating, which can be extended to air conditioning, budget permitting of course. In the garden he has worked out an irrigation system for the orange grove. We will be able to water the trees at the press of a button, or in a more economic version, by the turn of a tap. And given the high percentage of lime scale in the water we get from the well, we will need a water purifier to attach to the pump. All this is most impressive, and it's a relief to know the vital equipment is covered by the revised preventivo. Still, I have a complaint. According to Signor Crescente's plan, the garden is about to fill up with jumbo technical equipment. We already have a large cement bump of a well in the far right hand corner. Alongside the beautiful tufa wall there will now have to be a cistern, a filter, and a giant boiler. If we decide one day to have air conditioning, there will also be the requisite cooler. Outside the kitchen window is another boiler, a small one, but still a boiler. By the time they have finished, our erstwhile orange grove will look like a salesroom for hydraulic equipment. Surely we could take advantage of all that empty space in the warehouse and leave the garden free. The Husky dismisses the idea out of hand. The cistern must be near the well, and the boilers near the bathrooms and kitchens. "*È giusto?*" Of course, Silvio agrees, "*È logico.*" Gardens come well down on the scale of priorities here, as we have already discovered. But of course all this expensive equipment will have to be protected. Signor Cassano says he can build an extension of the wall in tufa to house the pump, and then we could grow climbing plants up the sides. The boilers could be concealed in a wrought-iron construction with a tiled roof. If the wisteria had survived, I think to myself, it could have been trained to wrap itself around the boiler shed. Perhaps this is the moment to raise a sore point.

"Oh, and Signor Cassano?"

"Si, Signora Caro*line.*"

"If you are thinking of cutting down any more plants in the garden, would you mind asking me first?"

By the time we get back to Milan, Marcello has come to terms with the revised budget, and we are both on a high in anticipation of our next trip to Sicily. The whole business is so absorbing that we keep asking ourselves why we didn't do it before. We congratulate ourselves on our new resolute attitude. And this time there has been no turning back. As summer approaches, our thoughts turn southward and frustration mounts at being so far away from where the action is. We are in daily contact with Silvio, who is now bearing the brunt of the stress involved in trying to get work done fast in Sicily. He has reminded us that we have a problem with *infissi*—doors and windows. I am somewhat taken back, as I thought the carpentry work was in the reliable hands of Signor Calandrino. The carpenter is a man of few words and extraordinary serenity. He never raises his voice but responds to any sort of aggression with a gentle smile. If you try to pin him down, his answer will be accompanied by a quiet chuckle, leaving you unsure whether he is serious or not. Apparently he has a nagging wife, so perhaps he has developed these tactics for the sake of domestic harmony. When we made our aborted start on the project three years before, he came to Santa Maria to give us an estimate. After a long and careful examination of all the doors and windows, he produced a hand-scrawled list of the work to be done. With Silvio's help we managed to decipher both the scrawl and the vocabulary, which was incomprehensible even to Marcello. The upshot was that he could repair twelve of the twenty-two doors and six of the eight shuttered windows. For the rest he would have to make new ones. Six months ago, when we finally gave him the go-ahead, he stood by his original *preventivo*. He had nine months to do the work. At the time that seemed plenty, and he accepted the commission with his usual low chuckle. We assumed he was well on his way through the work. Silvio puts us right.

"He hasn't even started yet. He's behind with everything because his assistant has left. He's working single-handed now. What do you want me to do?"

We are now in April, with only three months to repair so many infissi, and make twelve new ones, get them painted, fitted with glass, and installed. We are not talking about plain glass windows or standard doors. All the external windows and doors at Santa Maria come with internal shutters, and the two French windows into the garden also have Venetian blinds.

"Silvio, I don't know. Put the pressure on. Tell him we'll give the work to someone else."

"I doubt if we'd find anyone who could do all this work in such a short time. Anyway, I'll have a word with him."

Later on I hear the story from Massimo. He didn't know who Papà was talking to, but he could hear him shouting angrily and banging on the table. Stefi was so frightened that she shut herself in her room. Finally it seemed that Papà had succumbed to the Manzo Temper. When the study door opened, Silvio led a sheepish Signor Calandrino downstairs, across the courtyard, and into the car.

"How did Signor Calandrino take it?"

"He just kept on smiling."

Apparently Silvio had browbeaten him into committing for the repair work and then made him do the rounds of all his fellow carpenters in town till they found one who would make the new doors. All I get from Silvio at the time is, "Don't worry. *Tutto a posto*—it's all sorted."

THE LATEST NEWS FROM SANTA MARIA IS NOT GOOD. WORK has been held up for weeks, at first because Signor Crescente went missing, keeping Cassano's men waiting for him to lay the wiring, then because the graniglia floor tiles still haven't arrived. Signor Cassano blames Vito, who blames the manufacturer. And Signor Cassano has another gripe—the Spanish tiles we have ordered for the girls' mini-apartment. They are handmade and will need special grouting, which, he says, is not covered in the preventivo. Just to add to the Byzantine's lament is the problem that the ceramic sink I have chosen is not available in Sicily. He can't start on the built-in kitchen until the sink arrives. Vito is confident we will get it. He has networked all the sanitary ware distributors in southern Italy, and it seems they have finally tracked one down to a retail outlet outside Naples. But by now Signor Cassano is no longer sticking to his deadline. He is bitter about electricians, plumbers, and tile vendors in general.

"I build normal houses, hundreds of them, no problem, but this one is *molto molto strana*. It's giving me headaches. I'm losing money."

Meanwhile we are all anxiously awaiting the arrival of the vital sink. Apparently it is due to arrive at the docks of Palermo on July 15.

"O la Madonna," groans the Byzantine down the telephone, "the day of the *festinu*." Of course, the festival of Santa Rosalia. How could we have forgotten that everything, including the docks, shuts down in Palermo to celebrate the festinu of the patron saint? Vito promises to contact the depot as soon as the festivities are over. Three days later he calls us with the news.

"When they unloaded the sink at the docks, they dropped it and . . ."

"And?"

"Non c'è piu—it is no more." The dockers must have been in a hurry to get off and join the festinu. So much for Santa Rosalia—definitely *not* my favorite saint. Now who is going to break the news to Signor Cassano?

BY MID JUNE THE MILANESE SKY DONS A THICK DESIGNER CLOAK of smog in Armani gray, and the pollution count has climbed so high that it constitutes a health hazard. The mayor of Milan has made a public announcement advising parents to keep their children indoors, and the 100 percent humidity count makes any exertion comparable to weight lifting in a Turkish bath. Mosquitoes descend by early evening, trapping us indoors and putting our shady green terrace out of bounds. With the temperatures topping the hundreds, Mazara with its beaches and sea breezes beckons with all the allure of a Siren. Originally we intended to hold out in our Hades up north until the end of the month but with the arrival of the graniglia, the works at Santa Maria are gathering momentum at an alarming speed. Fed up with the Milanese summer, we convince ourselves that our presence on site is needed now, weeks before our scheduled summer holiday. I cancel my courses for the rest of the month, and Marcello decides to accompany me. He will stay at Santa Maria for a few days before returning to Milan to work out the rest of his sentence.

The sight of the house in the last stages of completion brings a rush of excitement.

Signor Cassano has pulled out the stops to get the graniglia tiles laid in time for our arrival. Even before they have been sanded down, it is clear that he and his team of *muratori* have done an excellent job. The rooms are not symmetrical and the walls are not even straight, but he has cut the tiles to create an exact fit. This is superb workmanship, and we make sure he knows it. The new floors are elegant and uplifting; the diagonal design of black and white squares open up an endless perspective, like something out of a Vermeer painting.

"Molto, molto ben fatto, Signor Cassano. Che bravo!" The Byzantine concedes an approximation of a smile.

"Well I had my doubts. I've never worked with these tiles before, but it has turned out very well. I can't guarantee the same for the Spanish stuff though."

Once the muratori have laid the floors in the bedrooms and bathrooms, all that remains for the Byzantine to complete are the kitchen and the fireplace. The latter has already been started. There is a large hole in the dining room wall, from which sprouts a giant metal funnel and tube, a sort of elephantine trunk that climbs the wall and disappears into the ceiling. Naked and unadorned, it serves as a constant reminder that the work is still in progress. Signor Cassano doesn't appreciate the hurry to get it finished.

"Who needs a fireplace in summer? You'll have it by Christmas, I promise."

As for the kitchen, Signor Cassano has solved the sink problem by getting the stonemason to make one from Carrara marble to match the worktops. It is huge, even bigger than the ceramic sink that met its end on the Palermo docks. He has constructed it so that the veins run diagonally to the join, creating a folded ink pattern effect. All the marble has been sanded and polished to a level worthy of the most intimidating of banks, a little grand for a sink perhaps, but very beautiful. The splendid blue gas cooker we ordered from Vito is installed, complete with a dashboard of shiny brass knobs and switches to play with. There may be no hob filter or cupboards yet, but we now have a working kitchen. We celebrate by borrowing Nanette's Moka machine and make coffee for Signor Cassano. Over coffee the Byzantine quizzes us about the housewarming party. Housewarming? I hadn't given it a thought. How can we have a housewarming if the house isn't finished?

"Well, before you move in, don't forget the *pesce fritto*."

Just what has fried fish got to do with a house renovation? I appeal to Marcello for help, which he is unable to provide. Aghast at our ignorance, our builder informs us that it is a very, very old Mazarese tradition to leave out a plate of fried fish and a glass of wine for the *patroneddu di casa*—the spirits of the house, personified by none other than geckos—house lizards—for the first night of occupation. Failure to do so would mean bad luck for the house and its occupants. When I ask Silvio about this later, he suggests the tradition is a leftover from the Romans' worship of their household gods, the Lares and Penates. This seems rather a grand precedent for a local custom, but we make a mental note to buy fish to fry before we move in. If there do happen to be such things as household gods lurking at Santa Maria, we certainly don't want to risk offending them.

Before returning to Milan, Marcello entrusts me with a few little tasks to complete, stressing their urgency with a stern *"Mi raccomando!"* One of them is to make sure we get electricity installed in our new establishment on the ground floor as soon as possible. In Italy, getting wired up is not just a matter of a phone call to the supplier. It involves a trip to the office of ENEL, the energy company, which, in Mazara, is open in the morning only and famous for its long lines. Before going I will have to collect a series of documents, including a copy of the building permit Silvio obtained for renovation when the villa was damaged by earth tremors in 1981. According to Silvio, this vital piece of paper disappeared long ago into the chaos of the *magazzino* and has now probably been recycled by rats. But not to worry. He thinks there will be a copy in the archives of the comune. Unfortunately, these have been partially destroyed by fire. But I am not to worry about this either, for he has a contact (*naturalmente*) who has promised to sort the whole thing out. In the meantime, he has indulged in a bit of raccomandazione and spoken to the *ingegnere*, the local ENEL manager, who happens to be a *compagno di scuola* of Marcello's. This way I will be received on appointment and avoid the customary wait. Nanette comes with me for moral support. Still unused to this Sicilian way of getting things done, I feel a pang of guilt when the ingegnere's door opens and he summons us into his office, leaving a crowd of people pushing for attention, thronging and thumping on the counter in a Sicilian version of a line. He begs us to take a seat—*"Prego, prego* Signora Manzo, Signora Manzo."

After asking us to convey his regards to the *Architetto* and the

Dottore (Silvio and Marcello), and making a polite enquiry about the health of Signora Maria, the *Ingegnere* asks me what sort of contract I want.

"What sort of contract do I want? What do you mean?"

"Well if you take a contract for three kilowatts, you will save on the fixed charges and pay a lower consumption rate. It does mean you won't be able to use more than one major appliance at a time though. If the washing machine is on, you can't use the hairdryer. When you do the ironing, make sure the grill is turned off, etcetera."

"What about the automatic pump? Wouldn't the system fuse every time a tap is turned on?"

"Well you'd have to turn off all appliances before you turned on a tap."

This does not sound like a family-friendly option to me. Against Nanette's advice, I go for the luxury option of six kilowatts, which, of course, is twice as expensive.

"Signora, can I see your *certificato di residenza*?"

"We're not resident in Mazara. We live in Milan."

"So the house in Mazara is your second home?"

"Well no. We don't own our flat in Milan. Actually Santa Maria is our *prima casa*."

"Sorry, but if you can't prove you are resident here with a certificate from the comune, we can't offer you a resident's contract."

"And so?"

"You'll have to take out a contract for nonresident's second home, which will cost you approximately twice as much as the standard contract." I see. Even the electricity service has a political agenda to milk the rich, only they have got the wrong man this time. It is grossly unfair but there is nothing for it. If I want ENEL to link us up to the mains, I will have to sign as a multihome owner. He adds up the charges: six-kilowatt nonresidents' contract plus connection fee plus fixed charges for the first two months. I hand over a check for the equivalent of nearly seven hundred pounds. Nanette is aghast.

"All that money and you haven't even been connected yet!"

"Not only that, but once we are connected, we will be paying four times as much for our electricity as you pay upstairs." I ask the ingegnere when they will be coming round to connect us.

"Just as soon as you bring in the building permit."

On our way out I note that the atmosphere in the waiting room has heated up. Two men are shouting at the little man behind the counter, both complaining about overcharging on their bills.

As the ingegnere's door closes behind us, one of them calls out, "How do I get to speak to the ingegnere?" I refrain from telling him it helps if you've been at school together.

CAROLINE, MARCELLO, THE FLOWERS, AND A FRIEND

TEN

La Famiglia

BEFORE I EVEN GET OUT OF BED THIS MORNING I CAN BREATHE
the heat. By now of course I should know that every day is going to
blast out hot air with the energy of a powerful hairdryer, yet once again
the sheer force of it takes me by surprise. While only yesterday Santa
Maria was teeming with activity, now not only have Cassano's team put
down tools and left, but also Silvio's family have gone on holiday. Lul
has moved in upstairs to look after her grandmother, but both are asleep
at this hour of the morning. I make my way down to the courtyard,
where Pino, the painter, is finishing off the last few windows. Without
the muratori for company, he works in unaccustomed silence. Meanwhile I have to get down to cleaning and sorting in preparation for the
arrival of Marcello and the Flowers on an evening flight from Milan.
Before I can start work, the dogs must be seen to. Unfortunately Chris,
the springer spaniel, complicated Silvio and Nanette's departure by
going into heat for the first time, just as they were about to leave. This
means she can no longer share the courtyard or come on walks with
the other two dogs, unless she is to be the object of Jack's unwelcome
attentions. So poor Chris has been condemned to solitary confinement
on the upper terrace, locked in behind massive wrought-iron gates to
protect her virginity and prevent her pushing through the balustrades
to certain death. First she must be fed and watered and her kennel

cleaned, and then I must do the same for Lucy and Jack and take them for a walk. Today my elderly fox terrier is not interested in going out. She has settled down in her favorite spot in the courtyard, where she can stretch out and bask in the morning sun. Jack has other ideas. He is fundamentally a city dog, not keen on aimlessly running about on his own, not that he ever does much running anyway. He must be taken out and walked properly. He will leave me no peace until he's had his after-breakfast constitutional.

I draw back the bolt of the courtyard doors to let Jack out. It is one of those glorious Sicilian summer mornings when the sky is doing its best to beat its own record for blueness and the air positively shimmers. I take a deep breath to drink it in. The mesmerising heat has cast an eerie spell over Santa Maria. There's not a sound to be heard except for the constant zing of cicadas and occasionally the low rumble of a car passing in the distance. The front garden is a great mass of purples, reds, and pinks, unkempt maybe, but a gorgeous feast for the eye. We head left for the *pineta*, the cool shady garden of pines beyond the parco. Jack trots along happily ahead of me, stopping only to snap at a butterfly or investigate the rustle of a lizard in the bushes. As we trample the dry pine needles underfoot, he collects various nasty prickles and burrs in his coat and paws. I feel I should kneel down, like Saint Jerome with the lion, and patiently extract them, but my recent hip surgery makes kneeling difficult right now, and I decide to postpone the operation until we get back. Fortunately, the lion is not complaining.

We make our way past the confines of the Santa Maria property and into the avenue of monumental cypress trees that cuts through the vineyards and links Santa Maria to the Maskaro tower. This is all that is left of the wood for which Santa Maria was once famous. Less than a hundred years ago the woods in these parts were so thick and shady that they were proverbial. The supply of wood here was thought to be as infinite as the tea of China, so that to burn all the wood of Santa Maria was a local expression for doing the impossible. Since then, sys-

tematic deforestation has taken place all over the island. Inevitably in a
society where the peasant had nothing but what he could extract from
the soil, trees would be cut for firewood and common land would be
taken over for cultivation. Now, with the exception of the sumac, the
eucalyptus, and the occasional lofty date palm, the pines and cypresses
of Santa Maria are the only tall-standing trees on the landscape for
miles around.

The Torre di Maskaro stands clear of the avenue of trees. It was
built on this flattest of terrains as a lookout point, and the view from
the terrace extends as far as Mazara and the sea. Since Silvio restored
it, it is one of the best preserved examples of a sixteenth-century
watchtower in Sicily. The four corbels on each corner of the square
roof terrace, although unusual for Sicilian architecture of the time, are
replicated in towers built by the Hospitallers of Saint John in Rhodes.
This has convinced Silvio of a link between this order of militant
knights and Santa Maria. It could be that the baglio was a comman-
dery for the Knights once they had been driven from Rhodes and
transferred to Malta. In this case they would have built the tower in
the Rhodian style, with elaborate decorations on the four corners of
the roof. Once it was a link in a chain of *castidruzze*—local towers
whose object was to protect grain. A post on horseback system carried alert messages from the coastal towers to these country neighbors. Over the door of the Torre Maskaro is a Latin inscription with
the date 1584 and the name of the architect. Funds permitting, Silvio
hopes to convert it to a studio one day, but in the meantime vandals
have been at work. There have been numerous break-ins resulting in
damage to the stonework and broken windows. Now the door is
locked and the key in Silvio's custody. If I had access, I would climb
the iron staircase to the terrace to get the best view of the countryside
north of Mazara, to gaze on what Tomasi di Lampedusa calls "the
immemorial silence of pastoral Sicily far from everything in space,
and still more so in time."

PALERMO AIRPORT HAS BEEN REBUILT RECENTLY AND GIVEN A mouthful of a new name—Falcone e Borsellino—in honor of the two Sicilian judges assassinated by the Mafia. The old name Punta Raisi, Fisherman's Point, was an apt reminder of the airport's juxtaposition to the sea. With the runway lying sandwiched between a rocky coast and a semicircle of perpendicular mountains, landing here was once a risky enterprise, which put Palermo well up on the list of Europe's dangerous airstrips. Fortunately since then, the runway has been shifted inland and enlarged. The flight Marcello and the girls are arriving on appears to be on time. I have to fight my way through the crowd packed into the arrival area. Pushing and shoving here is not accompanied by the wimpish *'permesso?'* or the apologetic *'scusi.'* No one respects his neighbor's space. Jostled mercilessly by jutting bosoms, boulderlike backpacks, and raised elbows clasping mobiles to big hair, I decide to retire from the fray and hope that Marcello and the girls will find me lurking at the back of the crowd.

Clem is first out, more beautiful than ever, and already tanned from a week in Crete with her friends to celebrate leaving school. Yeah, it was really cool. No she didn't go to Knossos, and sorry she didn't phone at all, but there was no phone in the village, etc. She starts telling me the saga of the previous night in Milan, how the three of them have spent a *notte in bianco*—a night in white, Italian for sleepless. Sleepless because of an alarm that went off in the building. Papà called the police. The police came to interview them. They had to wake the porter. The porter called the owner, who lived in Bergamo, an hour's drive away, and they all had to wait till four in the morning till this woman arrived to turn it off. Can you believe it? So no one got any sleep last night and they're all complete wrecks. And guess what? The airline tried to bump Melissa off the flight because of overbooking?

"So?"

"Well Papà had a go at the Alitalia chap (the Manzo Temper again), and yes it was terribly embarrassing, but it worked and here we are. How are you Mum? How's Nonna?"

Melissa emerges looking somewhat frail. She has just completed a month of internship for a commercial law firm in Milan. The "white night" on top of an uncomfortable spell in Milan in July, with its heat, humidity, and mosquitoes have taken their toll on my twenty-one year-old. She is too tired to say much and sleeps all the way to Mazara.

As we speed along the empty motorway flanked by mile after mile of wild eucalyptus and pink oleander, I remind Marcello that this is the first time in our lives we are taking our daughters home, our home, that is, not to a rented house or flat. He doesn't need reminding. Like me he is only too aware that we have inflicted a rootless peripatetic lifestyle on our children.

"Do you remember that time when they didn't want to be Sicilian?" Of course I do. When we first came to live in Italy as a family, the Flowers, then aged six and three, discovered they had an identity problem. Their smart Milanese schoolmates made it quite clear to them that it was acceptable to be half English, but to be half Sicilian—well, that provoked a few sniggers. Sicily was where the Mafia came from. Having a Sicilian father classed them as *meridionali*—rough immigrants from the south, who spoke a guttural dialect and inhabited squalid tenements on the outskirts of the city. Melissa and Clemency, who until this point in their short lives had had no problems with their Sicilian background, were so upset at these taunts that they opted for a change of status.

"Mummy," Melissa pleaded, "If we have to be Italian, can't we be Milanese?"

"Please," wailed Clem, "Why do we have to be Sicilian?" Unwittingly the two little girls had stumbled across the eternal Sicilian quest for identity, the heritage of a proud people who have been often invaded and just as often held in contempt. No longer preoccupied with playground survival, over the years our daughters have come to revel in their sicilitudine, even though our links with Sicily have been tenuous, based on summer visits as guests of Maria, Silvio, and Nanette. Now all that will change. Finally we are to have a Sicilian home of our own.

FAR FROM HAVING THE HOUSE READY AND FURNISHED FOR their arrival, I now have to enlist Marcello and the Flowers to help me effect the move. Meanwhile we all have to sleep upstairs. Before we can move in, the graniglia and cotto floors have to be treated, and the only person on the island qualified to give our floors the treatment they deserve, according to Vito, is a certain Signor Cannao, who hails from Messina on the other side of the island. Like our Husky plumber, he plays hard to get. When we finally manage to pin him down to a day he tells us he will have to make two visits, as the cotto needs to be treated twice with a twenty-four-hour interval to allow the first application to dry. This will take us up to the eve of the arrival of the first batch of guests. The alternative is to wait until after Ferragosto, not an option as far as I am concerned. Who knows how long we will have to wait if we put this off? Then there would be the added hassle of moving the furniture twice. We say yes. He says, "Right, see you Thursday morning at nine."

"At nine? It is at least a four hour drive from Messina to Mazara."

"Yes, I'll be up at four and on my way at five. See you Thursday at nine." By midweek the man from ENEL has come around and wired us up to the mains. Silvio's friend in the comune must have found the missing file. However, we cannot move the electric appliances into the kitchen until the cotto floors have been treated. Everything is on hold until Thursday. Signor Cannao is true to his word. He turns up with his team at nine as promised, but there is an unexpected complication. There is no electricity this morning. ENEL have decided to treat us to our first power cut. Signor Cannao unloads his machines and his magic buckets, surveys the house, then stands back with his arms folded.

"We'll just have to wait then, won't we? If necessary we'll stay the night in Mazara and finish the first application tomorrow." In other circumstances there would be no problem, but we have a very tight schedule. A phone call to the ENEL office brings some hope. The men are working on a fault in our area, which shouldn't take long to fix. This is promising, but we would like more precise information. We all stand around making what starts off as polite conversation but somehow deteriorates into a senseless dispute of the East versus West variety. Signor Cannao has definite ideas on the subject.

"Messina is very different from Mazara. It's a very modern town."

"Yes, of course. It was entirely rebuilt after the great earthquake, wasn't it?"

"And everything works too."

"What, you mean no power cuts?"

"Definitely not. Well, what do you expect? Mazara's practically in Africa, no, eh?"

Before Marcello can bring up the subject of "*Sicilia babba*," I have produced coffee and offer it around in plastic cups. The cigarette butts start to pile up. (It is considered acceptable to drop them on the floor as the house is still officially a *cantiere*—a building site, which means you can make as much mess as you like.) Signor Cannao sits smugly waiting. Marcello hits on an excuse to make a getaway from the tension that is building up.

"There's no point sitting around. Let's go and find out what these ENEL people are up to."

I follow him to the car and we set off to search for the offending electricity pole. It doesn't take us long to find it. At a hundred meters from our front gate, next to the local rubbish dump, there is an ENEL van. One man is on a ladder, working on the wires, while another two look on. As soon as the man on the ladder spots us, he starts to climb down.

"No don't come down," Marcello shouts up at him, "We are from that villa over there. We just wanted to know how long this is going to take." The man continues the descent. Now he is looking at Marcello face-to-face.

"I can't actually say. It'll take an engineer to fix it. We're just off to lunch and then we'll come back with a colleague."

"What about the other way round? What about getting your colleague here first? Call him on the phone and then go off to lunch. That way you won't have to come back."

Within half an hour the electricity is back and Signor Cannao is applying his special formula to our kitchen floor. There's a lot to be said for Marcello's hands-on approach.

All goes well once the job has started. The graniglia floors are ready after a single application. We can move furniture into those rooms.

"But whatever you do, don't set foot on the cotto." In other words the mini-flat and the kitchen are temporarily out of bounds.

"It's very important. See you on Monday." I pass on his instructions to Pino, the painter, and resign myself to the fact that we'll have to stay upstairs for a few more days. We get the weekend off to relax with no workmen in the house except Pino finishing off the last windows. Marcello and the girls want to join their cousins at the beach. What about moving the furniture? I remind them their guests arrive in a few days. Perhaps Pino could help us. But Marcello has identified another untapped area of human resources. There are twelve people sitting around Next Door, including six adult males. Let's get them over here. He shouts for Pilli over the fence.

Although Next Door has been filling up over the last week, I have seen very little of Pilli's family during the day, except for Maria Antonietta, the only one whose waking hours roughly coincide with mine. Her bedroom is adjacent to our kitchen and we chat over the fence when she comes out onto the terrace to water her geraniums. The rest of the family and guests usually asleep until midday, after which they emerge, one by one in beach gear, sip espressos on the terrace, and disappear in convoy to the beach to soak up the sun for the rest of the day.

We are in luck. The exodus to the beach has not begun, and Marcello manages to intercept the "follow the sun" program. All hands, if not yet out of bed, are potentially on deck. Pilli, of course, as Paterfa-milias and high blood pressure sufferer is exempted from any exertion. Goffredino, the eldest son and self-proclaimed hypochondriac, has one of the headaches for which he is famous. Eros, the youngest, is still in bed, which leaves Emiliano and his girlfriend, Michela, and Luca, Diana's husband, as the only suitable candidates. There is also a brother of Luca's, but he has come to Sicily to recover from the breakdown of his marriage and is treated with the courtesy due to a guest and the consideration due to a victim of female treachery. In the end it is the ever-obliging Emiliano and Michela who do the bulk of the work, with Marcello directing them, and Luca providing jokes and encour-agement. Maria Antonietta and Diana brew coffee in the Moka machine and pass it in plastic cups through the fence.

Meanwhile the Flowers and I have boxes of old kitchen utensils, crockery, and pans to unpack and wash, much of it littered with droppings

of cockroaches and rats. These are the rejects of my old kitchen in Brazil. Bent aluminium pans, countless mildewed wooden spoons, a motley collection of plates and glasses. How did I ever accumulate so much junk? Where are we going to put it all? Our initial enthusiasm soon wears thin, and in the end much of it is stuffed back into the boxes it came out of, which are stacked on the laundry floor. The ban on entering the kitchen gives us the excuse we need, and we flop onto the cane chairs in the sitting room just in time to catch sight of Pino barging into the kitchen. I have explained very carefully to him about Signor Cannao's special treatment of the cotto tiles, rounding off my instructions, Marcello style, with a firm *"Mi raccomando!"* But he knows better than to take notice of my nagging. There he is hanging the newly painted window on its frame, his large Nikes firmly planted on the freshly treated cotto. It's too late to stop him now. What comes in, must go out. I watch helplessly as the single trail becomes a double trail of Nike footprints clearly embedded in the surface of the terra-cotta floor.

As for the furniture-moving operation, the initial frenzy of activity soon expires. The helpers drop like ninepins while the sun pursues its pitiless arc across the summer sky. Even Emiliano has given in to the heat, and his tall wiry frame wilts visibly under the midday sun. He sits sprawled on the newly installed sofa, mopping a grimy brow with the sleeve of his T-shirt, his dreadlocks limp with sweat. Now that we have beds and chairs to sit on, Marcello as Project Manager decides that work can come to a halt. Some of the team is impatient at the late start to their day at the beach. They are anxious to establish themselves in prime positions before the afternoon crowd arrives. Every year there is a new "in" spot to develop your tan. This year it is a small rocky bay near the village of Torretta, called *Torre dei Saraceni*, after the lookout tower that serves as a reminder of the long-standing fear of Muslim pirates. Luca enthuses about the excellence of the on-site trattoria, which goes by the unpretentious name of *"il chiosco*—the kiosk."

"Today they've promised to make swordfish rolls. And you mustn't miss the antipasti. They're simply out of this world." Marcello and the girls have had enough hard work for one day, while the thought of fresh seafood salad has undermined my never-firm resolution to stay at home with the packing cases. This will be my first day at the beach this season. I am hesitant at the thought of exposing white flesh to the inhabitants of Next Door, all owners of carefully nurtured tans, except Luca's brother, who has just arrived and is almost as pallid as I am. My resistance is token, and we all decide to join them at the Saracen tower.

Eros is there to greet us, having surreptitiously slipped out during the furniture-heaving operation. Eros the Younger (namesake of Zio Eros Manzo), shamelessly spoiled from birth, combines great charm with a madcap reputation. Inseparable from his guitar, with his long black hair, curl of lip, and his way with girls, he could be Sicily's answer to Elvis Presley. He is hand in hand with an extraordinarily beautiful girlfriend. Together they are a His and Her outfit and even their names match: Eros and Erica. A perfect couple. She is shining and wet, like a sea nymph, with her long limbs, honey blonde hair, and luminous green eyes. I am beginning to regret my decision to bare the flab, but it's too late to cover up now. The tune of the season, a Mambo number, drives the point home as it blares out loud and clear from the jukebox.

"A little bit of Erica, on ma mind. A little bit of . . ."

At the Saracen Tower there is no sand where we can spread our towels, only flat slabs of rock worn smooth by the sea. The enterprising family who run the improvised trattoria have taken over the area, providing beach facilities for their patrons. Deck chairs suddenly appear for us, and a large green-and-white-striped parasol. No charge, we're told. It's on the house, the house being the *chiosco*, where we're going to have lunch. Melissa and I flop into the chairs gratefully. I notice we are the only people to have brought books. We open them but don't actually get around to reading, as the surroundings are too distracting. The novels resting on our laps turn out to be conversation pieces. What's it about? Is it difficult? How lucky you are to be able to read in English! We give up trying to read and step gingerly over the slippery rocks, looking for patches of sand among the algae. Where we stand in the shallows the water is no more than ankle deep, but there are already shoals of tiny fish. The water is crystal clear but freezing cold. The next step would be to leave the shallows and launch out beyond, where the shelf drops sharply and the water is suddenly deep. Nobody has plunged in yet, except the boys who have come prepared with wet suits. They go diving for sea urchins while the girls stand in a huddle, comparing the various shades of bronze of their sleek bodies. Melissa does not join in. In spite of a sunbathing session with Lucy in the cortile, she is still luminous white, but with pink splodges in places where the SPF 30 did not reach. Maria Antonietta and Diana wade through the shallows, leaning forward to splash themselves with the icy water. Melissa wants to know why the water of Mazara's beaches is always so freezing cold. From Tonnarella, the sandy beach north of the city to where we are on the south side, it's the same story, but if you go farther in either direction, the water turns warm again. A cold current from Africa is the usual answer, one that aims for this particular stretch of coast. Whatever the reason, it turns actual swimming into an exercise in masochism. We stand there until we are roasting before taking the plunge.

Back on dry land we interrupt a recipe conversation in progress, with Pilli's voice rising over other male voices. They seem to be arguing about a pasta recipe, the controversial issue being whether a certain dish should be made with bucatini (long, thick macaroni) or spaghetti. As every Italian knows, the choice of pasta shape is of vital importance to the success of a recipe. Spaghetti is a common favorite for a tomato-based sauce, but for this particular recipe, which is extremely rich, Pilli argues for bucatini, which are thicker and hollow inside, better for mopping up the sauce. Once this is settled, they go on to discuss whether or not anchovies should be included. Pilli insists the anchovy version is unorthodox; Lallo is sticking up for the anchovies. I pull my diary out of my beach bag and start writing down the recipe for *Bucatini alla fantasia* (name improvised by Pilli on the spur of the moment). Sicilian fresh tomatoes, pizzutelli (the small powerful ones with the thick skin), black olives, aubergines, a couple of garlic cloves, a generous dose of grated pecorino, and a dash of *peperoncino*. I also add anchovies.

"They're optional, so put *facoltativo* in brackets," Pilli insists.

Luigi Emilio arrives with the swagger and authority of the second eldest cousin of the Vaccara tribe, and consequently the talk turns from food to more macho topics. There is a lot of arm waving, and voices are raised even higher than before, although everyone is standing within inches of each other. Marcello is shouting too, as though he needs to assert himself when with brothers and male cousins. Pilli and Luigi Emilio, who do most of the shouting, have no doubt been communicating on this level ever since they were knee-high. Looking round, I notice that everyone in sight except the *chioscho* family is connected in some way. There must be at least thirty people here, but every single one is a Vaccara, a Manzo, an in-law, or an in-law of an in-law. It's as though the clan has taken over the location and has put up an unwritten sign—Family Only.

The male group thins out, leaving Marcello and his two brothers with their cousins Luigi Emilio and Lallo all standing in their bathing trunks on the edge of the rock with their backs to the sea. The talk has turned personal. I can see them lined up in profile—five large middle-aged men (all but Pilli and Marcello are six feet tall) all sporting the *pancia*. Could they be doing what I think they are doing—comparing their physiques? Judging from the tensing of shoulders and sucking in of stomachs, this must be a contest of paunches. As voices rise, an argument gets under way: Silvio has the biggest. What rubbish! (loose English translation of exclamation that refers to the male member), he has a bigger frame than the others, so it's in proportion. Well on that basis, Pilli's is the biggest. Look, it projects out like a balloon over skinny legs. Lallo's has no distinctive shape to it; it's all floppy, not like his elder brother's solid plateau. Marcello's is simply pathetic. Call that a pancia? As for getting rid of the pancia, the idea does not seem to have entered the waiting room of any of their minds. Which is just as well for now, as there is a feast to look forward to.

Luca was right about the food. We are served with course after course of delicious seafood, spaghetti with mussels, baked sardines, swordfish rolls stuffed with breadcrumbs and pine nuts. The supply of dry white wine flows in abundance. The lunch goes on and on, and the bill is ridiculously small, which only adds to the general *allegria*. Only the young people return to the sea after lunch. The rest of us linger on at the tables under the shade. More friends and family join us, including two more Luigis. The talk turns to local politics. Pilli, Marcello, and Luigi Emilio all shout at the tops of their voices, while Lallo listens, philosophically puffing on his pipe. The women move off back to the deck chairs. Lallo's wife, Rosetta, invites everyone to dinner later in the week. The invitation is for the day our guests will arrive. By common consent Rosetta and her mother, the Signora Giuseppina, run neck and neck with Pilli and Maria Antonietta in the Best Cooks ranking, so we look forward to a splendid dinner to launch the party season. The real holiday of days at the beach and family parties has begun. With the arrival of the Flowers' friends, all three establishments at Santa Maria will be full. Good-bye to the stress of the building project and welcome if not to peace, at least to partying and pleasure.

SIGNOR CANNAO HAS FINALLY LEFT, BUT NOT WITHOUT A PART-ing shot about the damage to the kitchen floor. The floors he has treated with his magic buckets are very *delicati*, to be cleaned in future with water only, no detergents or bleach. "Mi raccomando."

Maria's nurse, Signora Antonia, comes to help me clean up before the guests arrive. She turns up with her teenage son, Gaspare.

"Can you give him some odd jobs to do? He would like to earn some pocket money."

I have my misgivings, as in my experience teenagers and house-work are an awkward combination, and the results are usually less than satisfactory. Antonietta has told me about her son's difficulties at school, which she attributes to *nervosismo*. In his shorts he looks like an over-

sized Just William, the naughty ten-year-old hero of Richmal Cromp-
ton's tales of English suburbia. But I am prepared to give him the benefit of the doubt. I hand him a cloth and a bucket and ask him to take
the cane chairs into the courtyard and wipe them down. He has to ask
his mother to interpret because "the Signora talks strange." Signora
Antonia then sets to work mopping down the floors. Gaspare attempts
to turn on the hose, pricks his finger on the bougainvillea bush, and
starts to cry. "Mamma! Mamma! Help, help!" This seems unusual
behavior for a boy of seventeen. Will he be all right? Can we leave him
to his own devices? His mother is obviously used to his ways,

"Don't worry, Signora. He's a little nervous, as he's never been here
before. *Ma è un bravo ragazzo.*" She promises to keep an eye on him, so
we leave them to it while we go out to do a massive shop in preparation for tomorrow's arrivals. When we return laden with our supermarket bags, Gaspare is sitting in the courtyard, gazing at the sky.

"Ciao, Gaspare, are you tired?" He says, no, he's just taking a break.
As Signora Antonia hears us approach, a cross between a cry and a
scream escapes her lips.

"Signora Caro*line*, there's been a *disgrazia!*" An internal alarm bell
rings, as the word *disgrazia* is usually reserved for real tragedy.

"Signora Antonia, what happened?" But I can see for myself. She is
on her knees, furiously rubbing away at what looks like an amorphous
red stain, while large globules of pink oil are spattered all over the yellow and white floor. Between sobs and frantic rubs, the truth comes
out. Gaspare finished cleaning the rattan chairs and decided to turn his
hand to something more ambitious—to polish a table with a special
pink concoction his mother had brought. As the table was too heavy to
move, he applied the polish on the spot, but unfortunately dropped the
bottle on the floor. In his confusion, he stepped in the lethal pink liquid
before running to his mother for instructions. From the spiral pattern
of fuchsia footprints extending the length of the sitting room, he
appears to have executed some kind of tribal dance in the process. Signora Antonia is as desperate as Lady Macbeth to get the damned spots

out. Detergent, alcohol, nail varnish remover, nothing will do the trick. Silvio is summoned to come down to have a look. He says he has a special product that he uses for cleaning guns. Massimo is sent to fetch it. Everybody hovers hopefully as I get on my bended knees and go rub-a-dub-dub. Nothing happens. What was pink is still pink. Marcello grabs the bottle, then Nanette, Massimo, and Stefi. Five bottoms in the air and five pairs of elbows work furiously at what is now know as *la macchia*, a word that can be translated as mark, stain, or blob. All three would apply right now. Silvio looks down on us with sympathy. We give up, as, after all our efforts, the blobs are as pink as ever. I ask him if the fuchsia spattered tiles could be lifted and replaced. The answer is a sideways glance, a raising of Vaccara eyebrows, a definitive downturn of the mustache, and an expletive *"Ma!"* In other words: don't even think about it.

Maybe the Mastercleaner from Messina would know what to do, I suggest. Marcello calls Signor Cannao, who is now on holiday in the far-flung island of Lipari. As Marcello relates the incident in an ever-rising crescendo (why do Sicilian men shout like this?), I visualize Signor Cannao reeling in disbelief. Marcello asks him if he has a professional remedy (he doesn't) and if, to his knowledge, this kind of accident has ever happened before (no, it hasn't). By now the *Messinese* has written us off as *incoscienti*—reckless—or just accident-prone. In any case we come from the wrong end of the island: Africa. So what do you expect? Just as well Marcello never mentioned *Sicilia babba*, I think to myself. Who's the *babbu* and who's the *spertu* now? It looks as though we will just have to accept the situation, pink blobs and all. I feel nostalgic for the ugly old ceramic floor. At least it was accident proof.

But it is not only the graniglia that is the worse for Gaspare's attentions. There is also a huge red splash all over the wall, a great livid birthmark of a macchia. Normally you could just paint over such a splash, but Francesca, our London supplier of exclusive hand-mixed paint, was very explicit on this point. To quote: "Francesca's lime wash paints, once applied, cannot be retouched." Any pretensions I may have had to interior decorating are now firmly in check. Silvio stands by for support, sympathetic, but righteous. He doesn't want to rub it in, as it were, but he does mutter something about *white* paint being more *conveniente* than expensive imported products, and of course, you can always paint over it. In other words, I told you so. Meanwhile the news of the pink oil spill spreads like, well, an oil spill. Within hours we have visitors. Oh God! I point out that the house is not ready to show. It's not at its best and neither are we. The Marcello Manzos are definitely not in show-off mood today. Rosetta's mother, the Signora Giuseppina, puts my mind at rest.

"We haven't come to see the house, *amore mio*. We've come to see *la macchia*."

SELINUNTE

ELEVEN

Selinunte

OUR GUEST OF HONOR THIS SUMMER WILL BE MELISSA'S FIDAN-zato, whose arrival we are all eagerly awaiting. This is the first time we have been told about a boyfriend, but Bernie's reputation has gone before him. He has just got his First in English at Oxford. Brainy Bernie. At first the name is unfamiliar to Sicilians. Some of the members of Next Door, whose English leaves much to be desired, take his name to be "Burning" and assume it is an ingiuria. When we put them right, they seem disappointed, as their version was a lot more promising than the name of a lackluster saint. Bernie and his New Zealand cousin, Damian, arrive at Palermo on the same day as Clem's friend, Tania, and just in time for Lallo and Rosetta's party. The two boys have been back-packing in the South of France and are looking forward to some rest and comfort. Tania's suitcase has gone astray, but fortunately she wears the same size ten as Clem, although at five foot ten, she is a good deal taller. We give them minutes to shower and change before bundling them all into the car again. The boys appear in crumpled shorts and white vests, identical to the outfits they arrived in, but presumably cleaner, and Tania looks ravishing in Clem's halter top and skirt, which, on her long-limbed frame, give new meaning to the word minimalist.

Lallo and Rosetta live in a bungalow in a dusty road of beach villas, aptly named Via Africa. Once a run-down beach house, surrounded by huge patches of bare earth, interspersed with unkempt cacti, it has become an oasis of green, a jewel of a Mediterranean garden, exuberant with the vibrant pinks and reds of hibiscus, bougainvillea, and lantana. Now is not the time to admire the garden, however. As we enter, the real eye-catcher is the spread. A huge trestle table is already laden with the multicourse feast, and Rosetta is ladling out steaming portions of pasta al forno, for which she and her mother, the Signora Giuseppina, are duly famous. Of all the stars in the galaxy of Sicilian cuisine, this is my favorite, but the multitude of courses to follow is already on display, a sage reminder not to fill up too soon. The choice is vast, and it is hard to fight a greedy instinct to try everything. Marcello makes a beeline for the barbecue, where Lallo, Aldo, and a cluster of Luigis are grilling stuffed squid over the coals. The party is already in full swing. It is one of those multigenerational occasions where everyone is included, from great-grandmothers to babes in arms, and most people are related by blood or by marriage to the other guests. This is the first big get-

together of the festive season that leads up to Ferragosto, in our case an opportunity to catch up with people we haven't seen for a year, or longer in many cases.

All the Vaccaras of Mazara have been invited. Luigi Emilio holds forth on some topic of which he has exclusive knowledge, some conspiracy theory or the latest corruption scandal, while his wife, Enza, proudly recounts the exploits of her two American grandchildren. She swaps stories with Franca, another Vaccara cousin, who is cradling her first grandchild. I like the enthusiasm of these young, active grandparents and the way they share the task of parenthood. Manlio, Franca's husband, is an *avvocato*, a defense lawyer, frequently required to defend clients accused of association with the Mafia. This stressful occupation has taken its physical toll, and though only in his fifties, he wears a pacemaker. Like Franca, he is besotted with the baby. With grandparents like these, who needs parents anyway? Come to think of it, the baby's parents are nowhere to be seen. Manlio explains. His grandson, Ernesto, is staying with them for the summer so their daughter, Raffaella, can have some time off. I check to see if Melissa and Clem are listening to this conversation. Thankfully, they are not. I wouldn't like them to make any assumptions about their mother's role in their future families.

How different are the Sicilians from the English, and never more so than when they are enjoying themselves. It goes without saying that the main point of the evening is food (certainly not drink), with socializing coming a close second. All formalities are absent from this party. There is no effort to mix the sexes or the generations or to introduce people to anyone they may not have met, and no obligation to circulate, split married couples, or separate the children from the adults. All the social rules that I grew up with are possibly unknown, or more probably, ignored. The men stand together in a huddle, having what sounds like an argument, but one cannot judge the mood of an exchange by the pitch of the voices. Most of the women are sitting gossiping in a circle, with one eye on the smaller children or grandchildren running riot between the chairs. Rosetta and the Signora Giuseppina emerge from the kitchen at intervals, each time bearing fresh supplies. They pause to ask the guests if they have tried all the antipasti or if they would like another plate of pasta. They are not concerned that the party should go with a swing. It will anyway. What is important is that everyone should enjoy the feast. The rest will take care of itself. Every now and then I check out our young guests for signs of disorientation, but my fears turn out to be unfounded. The heat, the unfamiliar food, the immersion in a crowd of strangers speaking another language leaves them unfazed. Tania is perched on a stool, happily digging into a plate piled high with pasta and nonchalantly displaying her kilometric legs. Eros and his cronies are entertaining her and Clem in pidgin English. No worries on that score. Melissa has dutifully introduced Bernie and Damian to everyone, although lack of a common language between the locals and the new arrivals has reduced the exchange to sign language. The Signora Giuseppina is the most effusive.

"*Amore mio, il tuo fidanzato e bellissimo*. How happy I am for you! When are you getting married?" Melissa doesn't need to translate this time. Bernie has got the main message, but fortunately not the question about the wedding date. He beams back at Rosetta's mother, who already has tears in her eyes. Melissa moves him on before things get too emotional. Introductions over, they can sit back, enjoy the food and wine, and just talk. As usual, most of the guests give the wine far less attention than it deserves. The exception is Damian, who at first makes polite trips to the wine trolley, which become ever more frequent. Finally he takes the logical step of transferring a carafe to his table. No doubt this contributes to the feeling of *benessere* that takes over him. While Melissa and Bernie converse in whispers, Damian sits back wreathed in smiles. The Sicilian experience is a far cry from the small town life he leads back in Rotorua. He cannot contain his enthusiasm for the hospitality, the exquisite food, the exotic garden, and the strobe-like lights of the Mediterranean sky at night. He is enchanted with everything and everybody, even though he can understand very little of what they are saying. It is not long before he has passed around his

home address and phone number in a general invitation to visit him
down under.

With Santa Maria being so far from town and with only one car
between us, I had been wondering how we were to entertain a group
of Young Ones for two weeks. I need not have worried, for our young
guests could not be less demanding. Like most of their generation, they
tend not to surface until midday. Breakfast is a cup of tea and yogurt, or
a peach devoured on the hoof. Melissa and Bernie spend their time
reading together under the orange trees, Bernie engrossed in Eliza-
bethan poetry and Melissa in some obscure work of Japanese literature.
This unusual, monastic behavior draws admiring looks from Stefi and
Massimo, who have a habit of appearing at the upstairs window to
check out the Cugini and their guests. Clem and Tania make themselves scarce, either retiring to the Flowers' mini-flat for private girlie
talks, or wandering about the grounds to pick peaches, apricots, and
mulberries. I feel sorry for Damian, who is too discreet to hang around
with the fidanzati, but is not invited on the girls' walks. Clem has taken
a dislike to him and has no scruples about showing it. Her antagonism
amuses Damian, which only infuriates her further and provokes even
more sarcasm. But Damian is as serene as they come and quite impervi-
ous to teenage barbs. The only early riser in the group, he chats to me
in the kitchen until the others emerge. After breakfast he disappears to
some distant point in the grounds to meditate and, as he doesn't believe
in timekeeping, he tends to forget to come back for lunch; more often
than not I have to dispatch Melissa and Bernie to search for him.

After days of lounging and lotus eating the subject of culture
comes up.

Damian asks, "Aren't there supposed to be some Greek temples
around here?"

Clem raises her eyes to heaven, not merely out of disrespect for
Damian, but because she has a strong aversion to cultural excursions of
any kind. Still, we can hardly let our guests leave Sicily without having
seen a Greek temple, and this time noblesse oblige will bring Clem

along too. To answer Damian's question, I tell him there is indeed a choice of Greek temples in Sicily, of which the nearest are at Selinunte. Since it is no one's idea of fun to drive across the island in the torrid heat of August for a midday tramp around Syracuse or Agrigento (only fools and foreigners would attempt serious sightseeing of this kind in high summer), we are fortunate to have Selinunte on our doorstep. All the Greek sites in Sicily are open to the public right up until sundown, so we can set off in the late afternoon and explore the temples in the cool of the evening.

Although I have visited Selinunte more often than I can remember, I am always happy to return and wander about the giant columns of this abandoned city. On my very first visit to Sicily in the spring of 1975, Silvio and Vito took me there, and we picked bunches of poppies and *aipo*, the tough-stalked wild parsley, which the Greeks called *selinon* and which gives the city its name. I still have the photos Vito took of us striking poses among the fallen columns. When Silvio joined the Soprintendenza, he became responsible for the restoration of a sixteenth-century watchtower adjacent to the temple site. By that time he was intimate with the archaeology and history of Selinunte, and every time friends visited, we could draw on his encyclopedic knowledge. He played the part of the cicerone perfectly, and looked it too. In his white linens and Panama hat, he would lead the way from temple to temple, expounding at length on the niceties of Greek architecture, resorting to French when English failed him.

Silvio is not with us this time, so the guests will have to make do with me as a guide. I start by telling them Selinus was one of the great cities of the classical age. A sea power second only to Syracuse, it boasted four ports, one of which was Mazara. Founded around the middle of the seventh century BC, Selinus flourished as an independent city for less than two and a half centuries, only to be completely annihilated by the Carthaginians. . . . But as I suspected, they are not really listening and are probably happy to take the experience as it comes, without a lecture. Bunched up in the back of the car they would rather plug their ears into their music devices than listen to the story of the rise and fall of Selinunte. Cutting it short, I hand them a 3-D-pullout map of the ruins instead, which Damian and the girls ignore. Only Bernie gives it his polite attention.

"*Selinon*—parsley. It does seem a strange name to give a city."

"I agree, but it's not unique to name a city after a local plant. The Portuguese did call their greatest colony after pau-brasil, brazil wood."

"Yes, but that was a valuable commodity—not like parsley."

Bernie is right. In fact the Greeks considered parsley unlucky and even used it to induce abortion.

"There's a saying in English—parsley grows ripe in a cuckold's garden."

"What's a cuckold?" asks Marcello.

The approach to the site of Selinunte is uninspiring, with the usual collection of souvenir and ice-cream sellers lining a packed parking lot. Buses bedaubed in psychedelic colors spill out tourists of all shapes and sizes. After passing through a cement bunker to buy the entrance tickets, we join the serpent tail of people heading for the eastern hill. The sight that confronts the visitor on arrival is of huge fluted drums of tufa, heaped together with their capitals and roof supports, as though some capricious giantess had ripped off her necklace and flung it down in a fit of pique, causing the pieces to roll and scatter over the slope, before settling in untidy piles. Here and there she has tried to put the pieces together, but the general effect is one of devastation.

Spread over gently sloping hills and a wooded valley overlooking the sea, its idyllic surroundings make Selinunte the most picturesque site in Greek Sicily and we are fortunate that Silvio and his colleagues at the Soprintendenza have been successful in preventing the illegal building that has jeopardized other Greek sites on the island. The three great temples that once stood in line on the eastern hill belong to the fifth century, when Selinus was already rich and powerful. They have been labeled anonymously with the letters *E*, *F*, and *G*, and although it would be satisfying to know which deity each temple was dedicated to, the archaeologists are reluctant to hazard any but tentative opinions. Fortunately for us, temple E, possibly dedicated to Hera (Juno), was reconstructed in the 1950s. This is the temple visible from the sea. Bricks and new stone had to be used to supplement the original remains in the reconstruction, and this provoked such a hue and cry in academic circles that no more rebuilding was attempted. Bernie and Melissa agree with me that Selinunte would be less attractive to visitors if the temple of Hera were in the same supine state as the rest of the temples.

At Selinunte there has been a serious attempt to explain the ruins to the layman. The Soprintendenza have put up notices in four languages, but unfortunately the text they have produced is baffling. Unless you are an archaeology buff or happen to have been to school in Sicily, where children are made to recite the parts of a Greek temple by heart, you might be hard put to know your pronaos from your peristyle or your triglyph from your temenos. On our expedition the young people give up, content to scramble over the temple steps and wander between the soaring colonnades, taking photographs of each other. The New Zealanders are amused by the sheer vitality of a group of Italian teenagers, who shriek and hoot with laughter as they leap onto the stone drums, pretending to embrace a pillar or prop up a leaning architrave. Damian is amazed at the Italians' capacity to enjoy themselves. If he could understand what they are actually saying, he would be shocked. Translated into English the crude language might sound obscene to his discreet New Zealand ears. The Italians have one approach to cultural excursions. Damian has another. He strides off on his own to gaze from afar and do some solitary meditation.

The Young Ones like their culture in short doses. Having traipsed over the eastern hill and the acropolis, they have had enough of history for one day and are not keen on my proposal to trek another half mile to see the Phoenician altar to Malophoros, the pomegranate bearer. Clem in particular is in a hurry to go home. We are about to join the procession of camera-swinging tourists in shorts and flip-flops back to the parking lot when we realize that Damian is not with us. He was last seen an hour ago wandering over the northern hill. There is no question of getting lost here, but Damian does not wear a watch and has an Eastern disregard for the constraints of the clock. Clem turns accusingly to Bernie, "You'll have to go and find him. He's your cousin." It takes us nearly an hour to track him down. We find him sitting on the wall of an ancient Greek house, communing with nature. He is unapologetic about his disappearance and oblivious of any need but to contemplate the beauty that surrounds him; he invites us to join him to watch the sun as it turns the columns of Apollo's temple a glowing radiant gold, then takes its leave and sinks in a crimson haze of glory into the wine-dark sea.

THE ORANGE GROVE

Siculus Coquus et Sicula Mensa

"SCONGELATI!" ALL IT TAKES IS ONE BITE FOR PILLI TO DELIVER his verdict.

"These prawns were not bought fresh—they've been frozen!" He is standing on the other side of the fence tasting a plate of barbecued prawns we have passed over for his approval. Pilli's house, which he converted from a disused warehouse back in the eighties, lies at right angles to the villa itself. His terrace is separated from our citrus grove by shoulder-high railings and a row of cypress trees that have achieved a miraculous height of twenty-five feet or so in the ten years since they were planted as seedlings. There is no communicating door, an arrangement designed to ensure a privacy that now seems nothing more than a symbolic denial of what is actually communal living. We can see and hear each other, but we cannot get physically close. Unless we are prepared to use the ladders precariously propped against the railings, (certainly not for me) or take the long, circuitous walk through the grounds to reach their front entrance on the other side, we have to make do with this zoolike arrangement and communicate through bars. Apart from Pilli, Maria Antonietta, Lul, and the three boys, the household

Next Door includes Diana and Luca, their daughter, Onda, Luca's brother, and various fidanzate, not to mention any other friends who happen to be staying. The house is bursting at the seams, but this gregarious family has come up with an overspill solution. They have brought with them a Bedouin tent acquired at a flea market in Cairo, which Emiliano has rigged up on the roof and decorated with exotic fabrics and cushions. While it serves as a gathering place for the boys and their friends until the small hours of the night, it also doubles as an open dormitory for extra guests. One night we counted up to thirty heads in residence Next Door.

With numbers like these it is impossible for the three families at Santa Maria to eat together on a nightly basis. Instead we have devised a mutual tasting system, whereby we pass the product of our efforts at the stove, oven, or barbecue up and down the stairs to Silvio's establishment or over the fence to Pilli's, and receive their offerings in return. Upstairs the cooks are gifted but distracted. Next Door is a hive of dedicated supercooks, and since they are already catering for numbers that would challenge even the most experienced restaurateurs, most of the food sampling traffic is one way—from them to us. Tonight, however, it is our turn to send an offering over the fence. On a trip to the fish market this morning Marcello was seduced by a tray of giant prawns, which he brought home to grill on the barbecue as a special treat. We sprinkled them with the juice of a lemon plucked from our garden, sucked the flesh from the heads in true Sicilian style and ate them with our fingers. They were delicious; at least we thought so. Pilli, the acknowledged authority in this family of foodies, thinks otherwise.

"You've been taken for a ride." This does not go down well with Marcello, who prides himself both on his negotiating skills and his eye for quality, especially when it comes to buying fish. He has haggled over food in markets all over the world, from Bangkok to Buenos Aires and has always held his own. How come he has been palmed off with frozen prawns in his own backyard? I did notice at the time that the prawn vendors were in a great hurry to get rid of their wares, agreed to a hefty discount and threw in an extra half kilo as a gesture of goodwill. But at the time I put it down to the lateness of the hour, as serious shoppers have been and gone by nine. First thing the next day we are back at the fish market to settle the score. Our fish vendor and his lanky son are pleased to see us.

"Dottore! Signora! Today we have swordfish *freschissima!*" But Marcello is not interested in swordfish right now. He goes straight to the point. I am apprehensive, as always on these occasions, in case the Manzo Temper should erupt.

"So those prawns you sold me yesterday were frozen. Were they or weren't they?" There is a short pause. It looks as though they are about to deny it, but the old man confesses.

"Well, yes, they were put in ice on board the fishing trawler. But only ice, none of your nasty chemicals, and if the boat was out for long, they probably did freeze. That doesn't mean they were frozen, Dottore, only frozen on board."

"So you admit they were frozen! I don't come to the fish market of Mazara del Vallo—*Mazara del Vallo!*—to buy frozen prawns. I can go to any supermarket for that." Marcello makes a dramatic exit. I follow him sheepishly, relieved that there has not been an angry showdown. The old man shrugs his shoulders and goes back to cleaning fish, but his son runs after us and follows us to the car.

"Dottore, Signora. Don't you know that prawns are always frozen on board? *Sempre, sempre.* They would go off otherwise."

Marcello insists, "*Per me, congelato è congelato.* Frozen is frozen. Those prawns were not fresh. How do I know how long they had been defrosted?" We are already in the car and the poor young man is pleading with us to listen. As Marcello switches on the engine he grips the car door.

"Dottore, Dottore, *Ascolti!* Listen, please, you can ask anyone. They were only frozen on board." As we drive away he is still shouting after us "Frozen on board!"

Mazara del Vallo, the most important fishing port in Italy, is the mecca of pesce, what Naples is to pizza or Parma to *prosciutto*. With three hundred fifty fishing vessels bringing in their haul to Mazara, one might be forgiven for thinking that fish would be plentiful and relatively cheap; but one would be wrong. The fish market is an unimpressive collection of a dozen or so stalls housed in a cement bunker by the port. Where have all the fishes gone then, I ask Capitano Aldo, our expert on all things associated with the sea. Up north mainly, he explains, to the restaurants and supermarket chains of Rome, Milan, and Turin. About 90 percent of the catch is sold directly to wholesalers offshore and never reaches the local market, a convenient arrangement for *marinai*, which ensures that they bypass local taxes. Here in Mazara we just get the leftovers, and even these are no less pricey than anywhere else in Sicily. It is not surprising therefore that the fish that appear most frequently on the table are the cheap and plentiful *pesce azurra*—sardines and anchovies. These are Sicilian staple fare, and when it comes to fish, the people of the province of Trapani, which includes Mazara, are particularly enamoured of the small and the salty. Before I came to Sicily I had always removed any anchovies that found their way into a salade Niçoise or onto a pizza topping, but as soon as Maria introduced me to *pasta con le acciughe*, I became a fan of the salted anchovy. Once I had been won over, I was ready to discover the taste of fresh anchovies marinated in vinegar, a rare and delicate taste, and only a very distant relative of the strong salted version.

Just as I revised my opinion of anchovies, so too sardines acquired a new image. Not just confined to a tin to be opened for an emergency supper of sardines on toast, here sardines are eaten fresh. It took Sicilian ingenuity to teach me that they are delicious when rolled and filled with seasoned fresh breadcrumbs, the poor man's basic stuffing. They call them *sarde alla beccafico*, "sardines like songbirds," and the plump little rolls with their tails poking out really do look just like little warblers. Maria Antonietta has turned the presentation of this dish into a work of art. Each little beccafico nestles in a bed of bay leaves and between each morsel she inserts a thinly sliced disk of an orange. The combination of silver gray fish with deep green leaves and the brilliant semicircles of orange cry out for a Caravaggio to immortalize the sight. I compliment the cook on her work of art.

"Maria Antonietta, *sei un' artista.*"

"*Che si mangia 'sta sera?*" What's for dinner? is, as always, the first question of the day, just as it was a generation ago, when Maria and Goffredo presided at the breakfast meetings. Our guests have left, and we are planning dinner at Pilli's house for just the Manzos, plus Diana, Luca, and Onda. There will only be sixteen of us, a small gathering by Santa Maria standards. Pilli and Maria Antonietta have emerged from their ground floor bedroom onto their terrace, where Luca and Diana have brought plastic cups and a brimming Moka machine. We conduct our conference as usual through the fence in a gap between the cypress trees. Marcello has already been to town for the newspaper, which he hands to Luca for first perusal. Although Luca is from Bologna in Reggio Emilia, the region with the reputation for the best cuisine in Italy, he has established a role as Family intellectual (second only to Silvio), and he never takes part in the menu conferences. Marcello and I hang on to the railings, while the others recline on deck chairs, sipping coffee in their bathing suits. There is the usual postmortem on the dinner of the night before but the main talk is about the menu for tonight. We will have pasta of course, and it goes without saying that we must have fish. In spite of their names (*manzo* means "beef" in Italian, and *vaccaro* means "cowherd"), the Manzos and Vaccaras come from a culture where fish eating is bred in the bone. Meat comes in a very poor second to fish, and you get the impression that they eat beef as a source of protein, more from duty than for pleasure.

"*Allora, che si mangia*—so, what are we going to eat?" Maria Antonietta offers to take care of the pasta, her speciality, but as for the *secondo*—the main course—they suggest we keep it simple, which means a barbecue. Everyone else is off to spend the day at the beach at Torretta, so Marcello and I are to do the shopping.

"What shall we get?"

"Whatever looks good. Just use your judgment." I have learned my lesson by now. You don't set out with a preconceived idea; just take inspiration from what's on offer. If you want the best, you must take what is in season. This will be the first time we go to the fish market since the frozen prawns debacle, and this time, Marcello decides, we need some support. With experts Pilli and Aldo otherwise engaged, we are going to call on Rino. Rino and Marcello were *compagni di banco*, which means that their relationship goes back to when, as eleven-year-olds, they shared a desk at school. From Rino's account of his school-days, he must have been the Sicilian version of Tom Sawyer. Legendary for his escapades, he was expert at what he calls *fare il siciliano*—playing truant—which he covered up for by forging sick notes in his father's writing. He would persuade a friend to follow his example and organize a *spaghettata*, if the parents were out, or a barbecue on the beach. He met his punishment the day he came home having caught the sun and his father demanded to know just how he had managed to get sun-burned during a Latin exam.

At first the streetwise Rino was intent on teasing and torturing Marcello, the wimpish newcomer who had never been to school before, but gradually he turned from tormentor to protector. He has been a staunch friend ever since. It is not just the long-term friendship that binds Marcello to Rino, but the personality of the man himself. He is deeply attached to his friends, as most Sicilians are, but what distinguishes him from everyone else is his great capacity for kindness and generosity. He always has time for his friends and goes to extraordinary lengths to help them. People like this are of course popular, and Rino's people skills have provided him with a network of contacts that have made him an important figure in the community. He is one of the town's most popular *personaggi*. Today we find him presiding, as usual, in the office of the family furniture shop, with its vast showroom of damask sofas and imposing bedroom furniture in polished walnut, dimly lit by crystal chandeliers. The atmosphere is dark and grand in the Sicilian mode, very much the style that would have appealed to the Zia Checchina. Rino is huddled with a client, but he leaps up as we walk in. A stout, ungainly figure in a gray suit, with a *pancia* that would be described as *importante* even by Sicilian standards, he is all smiles, delighted to see us, as we knew he would be.

"Caro*line*! Marcel*lino*! (Marcello is still called by the diminutive in Mazara even though he is approaching sixty).Why haven't you come to see me before?" He grips Marcello hard on the shoulders, and they exchange the male Sicilian *bacio* on both cheeks.We explain about the work on the house and the constant stream of guests, but he is not satisfied. The truth is that he would like us to Get Close every day, and indeed, we say, there is nothing we would like more, if only we were simply on holiday. Marcello recounts the shaming incident of the frozen prawns. Rino interrupts.

"Marcellino.What did I tell you last time?" Marcello can't remember as Rino always has a lot of advice to give. But we are used to Rino's conversation technique, which follows the pattern of the Socratic dialogue with the difference that, having posed the question, Rino sometimes likes to supply the answer himself.

"This is what I told you.You are a forestiero in Mazara.You've been away for a long time.You don't have the right contacts any more, and you can't trust just anybody. I'll show you where to go. Leave it to me." He turns to shake hands with a waiting client, then pushes him in the direction of his assistant. Friendship comes before business.

"*Allora, andiamo.* It's time I took you shopping."

The first call was to be the fish market, but Rino wants to stop off on the way.

"Where do you get the best sausage in Mazara?" For years the Manzos have been going to Signor Vassallo in Via Castelvetrano, but we opt to keep quiet and wait for Rino's answer.

"This is where." He ushers us into an unprepossessing establishment on the road that leads to the port. The shop is practically empty; there is not a sausage in sight, only a selection of offal and a few cuts that are distinctly unfamiliar to me. The proprietor is huge and red-faced. He reminds me of the grinning butcher in a film we have just seen—the Mafia spoof musical "*Tano da morire* (*Love You to Bits*)," set in the Vucciria, the street market of Palermo. He greets Rino loudly, and for a second I almost expect him to burst into song like the Mafioso butcher in the film.

"*Wheeeh Senatore!*" he booms. They lapse into dialect so abstruse that even Marcello strains to follow them. For my benefit Rino reverts to Italian.

"Vincenzo was in our class at school. Do you remember him Marcellino?" Marcello obviously doesn't but attempts a convincing nod. Rino turns to the butcher, now in dialect again, with his arm around Marcello's shoulders,

"*Chissu è me frati. Capisti?* Vincenzo, I want you to know, Marcello is my brother. Is that understood? Which means when he comes here to buy beef, pork, sausage, or whatever, you treat him the way you treat me. Only the best for my brother, okay?" Vincenzo gets the message and stretches his grin wider still. He snaps his giant fingers, and a skinny Tunisian boy emerges from behind a plastic curtain.

"*Caffé per quattro.*" Rino nods approvingly and the boy darts across the road to the bar to come back with our espressos in cups with glass lids. Rino wants us to try Vincenzo's meat. I can't think of anything we would do with the hearts and livers on display but give them to Jack and Lucy. They are not to have this treat. As it turns out, Signor Vincenzo has something else in mind. He has ready-made *involtini alla siciliano*—veal rolls stuffed with breadcrumbs. They were prepared for another customer but he is going to let us have them instead. With Rino's eye fixed firmly on him, he refuses payment altogether. As we leave the shop, he holds out a massive hand and treats each of us to a mighty squeeze.

Once outside, Rino turns to Marcello.

"If you don't remember Vincenzo, you must remember his brother, Flavio. The two of them are very close, but Flavio has taken a different route. He's the consultant in neurology in a major hospital up north."

"One brother's a butcher and the other's a brain surgeon. Isn't that a bit strange?" Marcello seems surprised at the divergent careers of these two *compagni di scuola*.

"Yes, it does seem a bit strange; they did go into very different walks of life." Rino pauses for reflection. Suddenly hit by a brainwave, his face lights up.

"Well, if you think about it, it's not really strange; they're both very handy with a knife, so you could say they're in the same business, just different branches of it."

We decide to walk down to the fish market by the port at the mouth of the river. The Mazaro, once a navigable river and natural port, has gradually silted up over the centuries. In winter a tiny trickle of a stream still makes its way down from the rocks and grottoes on the outskirts of the city, where the early Christians hid from their Roman persecutors. Under the Arabs the surrounding area was one of lush vegetation and it is still known as *Miragghianu*—the Emir's gardens. Only one generation ago the Mazaro still flowed through gardens verdant with citrus and other fruit trees. Marcello and Rino remember going there as children on family outings to celebrate Pasquetta—the Easter Monday picnic. Now nothing grows but reeds and wild canes, and all that is left of the River Mazaro is the deep port canal, which has offered safe harbor for vessels ever since the Phoenicians first landed there.

We are now in the road that runs parallel to the docks, leading to the ancient Piazza del Bagno, and already there is an almost imperceptible difference in atmosphere from downtown Mazara. Male groups stand in defensive huddles outside the bars, arguing in an urgent, guttural Arabic. The men are shorter and darker, the facial expressions more intense, the voices muted. It is though we have already crossed the Sicilian Channel and found ourselves in a North African town. As we reach the port itself, the vista opens out onto a scene teeming with action. A motley assortment of boats is moored along the docks, each with a small cabin and deck littered with nets, ropes and fishing tackle. Most of them have seen better days, the paint faded and peeling and the hulls streaked with rust, but anything smart and new would be out of place here. The forays of these smaller boats are limited to short trips, mainly providing for the local market. The trawling fleet that makes

Mazara famous, the *pescherecci* that cross the Sicilian channel to the shores of Africa, or set out westward across the Mediterranean to cast their nets in the Atlantic, no longer dock here in the mouth of the Mazaro but in a new artificial harbor west of the old port, backing onto the long stretch of the Tonnarella beach. Dominated by the Capitaneria—the harbor office, this vast slab of concrete is a soulless place. A harbor without people, it provides docking space for the bulky pescherecci in the lull between voyages. Tackle is neatly stacked on deck, and the boats lie completely still on a sea as smooth as glass. The only sign of life are the seagulls perching on the statue of Saint Vitus, Mazara's patron saint, who guards the entrance to the port, accompanied by his dogs.

So it is not in the main harbor, but here at the mouth of the Mazaro where the action is. On the north bank is the arsenal with the constant hammering of blacksmiths and carpenters at work in the open. The opposite side is lined with open workshops for the dyeing of nets and the soldering of tin. Men with mustaches and cloth caps sit watching the younger men at work. Others stand arguing in groups while stray dogs recline in the shade of the awnings. An unmistakable port smell of fish, engine oil, and bilge water pervades.

A crowd is gathering around a fishing boat as a fisherman unloads his catch directly onto the dock. Sicilian housewives are bent over the nets, their solid frames blocking our view of his catch. The Tunisian fisherman is shouting a mixture of Sicilian and Arabic. The women shout back at him in dialect. The men and women must understand each other somehow, as the haggling continues unabated. A matron emerges triumphant with her shopper bulging with *triglie*—tiny red mullet with rosy red scales glinting in the sunlight.

As we wander along the dock, Rino strides ahead on his short legs, acknowledging the salutations that come from passersby. *"Buon Giorno, Dottore."* As we approach the market, one of the vendors calls out *"Onorevole, venga!"* Rino has never been a member of parliament, but the joke title confirms him as a man of status. Inside the covered market they all treat him respectfully, interrupting business to exchange a greeting.

Rino introduces Marcello to his chosen vendor, and I overhear the same phrase that I heard at the butcher shop: *"Chissu è me frati. Capisti?"* As the men haggle, I wander from stall to stall marveling at the shapes, sizes, and colors: the skate fish with its hideous elephantlike trunk, the piles of plump silvery sardines, an abundance of bream and sea bass. I am fascinated by the *pesce di San Pietro,* as round and flat as a pizza with long shaggy fins. This fish is what we call John Dory. But why do Italians call it after Saint Peter? Rino doesn't know, but later Aldo supplies the answer: apparently this was the fish that Saint Peter took the shekel coin from and the dark oval markings by the fins are the saint's fingerprints. Shall we get some of these for dinner? Before deciding, Rino interrogates the vendor; he wants to know exactly what time this fish came in. Rino declares he never eats fish that has been out of the sea for more than four hours. Not wishing ever to be put to the fresh fish test, I make a mental note not to serve fish next time Rino comes to dinner. Having passed their sell-by date by an hour or two, the Saint Peters do not pass muster, so we move on. Each stall has the humble fish essential for couscous: small groupers, eels, and blotchy red scorpion fish still gasping on the marble slabs. We toy with the idea of fish soup but without Pilli's authorization, we decide to play safe. Our choice will have to be top-quality fish, suitable for barbecue. The issue of frozen prawns comes up again.

"As for prawns, Marcellino, the truth is you will never get them fresh from the sea unless you catch them yourself. All the fishermen put them on ice on board, and then they are left to thaw out once they get to the market. The problem is you never know how long they have been unfrozen. If you want my advice, when it comes to prawns, forget the fish market. Come with me. I'm going to take you to meet another friend."

We follow the bank of the river past the old tuna station, a chain of warehouses that once belonged to Luigi Vaccara, some abandoned, and some converted into supermarket depots or garages. As the river sweeps off to the left, the road leads to a piazza flanked by a shiny new building bearing the name Marmoreo's Frozen Fish Emporium.

"This is where you buy your prawns from now on. Come and meet Signor Marmoreo."

One whole wall of the shop is decorated with a giant fresco. It depicts a tall, cadaverous-looking man, clean shaven, with a high forehead and long thin nose. He is standing over a slab of tuna with knife in hand, posing for the artist, while an assistant skins a fish on the side. The table is covered with a variety of Mediterranean fish, and the port is visible in the background, with the domes and campanili of Mazara appearing faintly on the horizon. The knife wielder, I presume, is Signor Marmoreo, a fish vendor with artistic leanings. I suspect he was inspired by the famous Greek vase from the Mandralisca Museum at Cefalù, with its illustration of the sturdy tuna vendor, who contemplates his wares with evident glee. The vendor in the fresco, unlike the one on the vase, has a lugubrious air, more redolent of a dour Aragonese *hidalgo* than the jovial Greek fishmonger of the Cefalù vase. There is no mistaking the owner of the shop when he appears in the flesh to welcome us. He is even more severe-looking than his portrait. As he shakes my hand, I cannot help thinking that this fish vendor with his long thin nose and cavernous cheeks, could have stepped straight out of a court of the Spanish Inquisition. As his appearance indicates, Signor Marmoreo is a man of few words, but Rino is expansive, particularly as the talk is of one of his favorite subjects—fish. The prawns on display are indeed magnificent. It's not just the size that is staggering, although each prawn is the length of a man's fist, but the color. Where have I seen prawns of this deep red hue, almost as dark as fresh blood? I remember now; it was in Aldo's courtyard at Torretta, when he grilled them on the barbecue and served them whole. It had never occurred to us that they had been frozen. I ask Signor Marmoreo and he smiles for the first time.

"Yes, the Capitano is a regular customer here."

"And the red color?"

"These are females still bearing eggs, which you must eat together with the rest. They are the Sicilian answer to caviar."

Having now changed our minds about frozen prawns, we buy them in a five-kilo box, beautifully packed in rows, like scarlet Christmas crackers. Before we leave, Signor Marmoreo has something to show us. He takes us to one of the walk-in refrigerators behind the shop. One room has been reserved for a special catch that came in just this morning. He opens the door and there on the floor lies the gigantic silver form of a blue-fin tuna, the largest fish I have ever seen in the flesh. Rino reckons she weighs as much as he does. A hundred and ten kilos 250 pounds) says Signor Marmoreo, so Rino is not exaggerating. The sight of the long rounded body of a powerful, athletic creature lying inert on the floor is sad enough to draw tears. There are no wounds on the body. At least she was spared the bloody battle with the speared fishing hook of the *mattanza*. But how did they catch a tuna in August when the season is only in May and June?

"She was on her way back from mating in the Black Sea. Very few of them survive to make this return journey. If they are not slaughtered in the mattanza on the way, the Japanese get them with their long lines. You are the first to see her as she only came in an hour ago. I have Japanese clients waiting for a catch like this. Would you like me to call you when I carve her up?" If I hadn't seen her lying there in all her splendor, I would have jumped at the chance of a fresh tuna steak or a sashimi, but the thought of taking even a mouthful of this beautiful creature is out of the question.

The educational shopping trip is not over yet.

"What about fruit and vegetables now?" We have stopped off at Rino's greengrocer, and he is about to make the introductions. I'm thinking to myself that the last thing we need is a trip to a *fruttivendolo*. I explain to Rino that we visited the greengrocers only yesterday and went home laden. We have a lot more eating to do until we need to stock up again. In any case, we already have a very satisfactory relationship with our greengrocer. He has been to great trouble to obtain yellow peppers for me, which are rarely available in Mazara, as the Mazaresi prefer the green variety. When I remarked once on the lack of

choice of salads available, he went all the way to Castelvetrano to procure rucola—not a Sicilian staple but a northern fad as he pointed out
with an indulgent smile. Rino listens to all this patiently.

"Yes, I know him. He's one of the best *fruttivendoli* in Mazara,
but . . . ," and somehow I know Rino is going to have the last word,
"Does he give you female vegetables?"

My biology may be a little rusty, but I thought fruit and vegetables
were the products of cross fertilization. Since when did sexual discrim-
ination affect vegetable shopping?

Rino explains, "The best vegetables are female because they have
fewer seeds. It is the seeds that make the flesh bitter. Take this melanzana
for example," and he picks up an aubergine of the giant round Tunisian
variety on display, big enough to feed a family of six.

"See where the stalk is? That's the head. Turn it upside down and
look at the markings at the other end. Those are the genitals. If the
mark is elongated, you've got a male, if it's nice and round, you've got a
female, and as I said, the females are better." He picks up the aubergines,
one by one, examining them like a doctor making a diagnosis.

"What did I tell you? These melanzane are all female. Didn't I say
that Rino Bocina only buys the best?" I can't wait to get home to
examine the aubergines I bought yesterday. Meanwhile Rino drives us
back to his shop to pick up the car.

"Marcel*lino* and Caro*line*, I have a present for you." He emerges
from his garage with a crate of bottles of red wine.

"This is the wine I've had bottled for my friends. No, of course I
don't grow my own vines. No time for that, but this is the next best
thing. I buy the must from my brother-in-law and have it processed and
bottled myself at a friend's *cantina*. This is a Merlot, not a sophisticated
wine, but it's a *vino genuino*. Now look at the label. My daughter
designed it. What does it say?"

The label on the bottle shows a couple sitting at a table with raised
glasses. Like Socrates, Rino already knows the answers to his questions,

but Marcello reads out the inscription obligingly: "For Marcello and Caroline, a wine to be drunk with friends."

"And remember you two. As long as you are in Mazara, from now on you don't buy wine. Whenever you need any, just come to me."

As it turns out, Rino's Merlot is undrinkable—a black potion that manages to combine the toxicity of Brazilian cachaça with the flavor of Ribena. Not long after the initial gift, Rino himself becomes aware of this and quietly suspends the donations. However we are touched by his kindness and the affection expressed in the special label. We invite him and his wife, Vera, over for coffee after lunch. Blonde and vivacious, fortunately Vera has the deep reserves of resilience needed to manage her husband's compulsive sociability. She is a retired schoolteacher, one of those very lucky people who took advantage of the Italian system that allowed state employees to retire on a full pension after only fifteen years' service. The system, now abolished, was called Baby Pension, a joke in her case, as she waited for retirement until she had her third child. Before they leave, I take the opportunity to ask Rino to make a diagnosis of our aubergines. I have two large specimens in my fridge, which I was going to use to make aubergine pizzas. Thickly sliced and fried in olive oil with a dab of fresh tomato sauce, a sliver of mozzarella and crowned with basil leaves, they will make a delicious vegetarian main dish. But first we must establish their gender.

Marcello and I have examined their nether regions, but we were unable to come to a conclusion. In both cases the shape is neither long and thin, nor fat and round. The markings on one are smooth and egg-like, on the other shapeless and puckered. Rino looks hard at them before delivering the verdict.

"This egg shape occurs sometimes and means that the vegetable is male with female leanings. Call it gay if you like. But as for this one, I've never seen one like it before. If you ask me, it's a hermaphrodite."

BELLY DANCER AT THE FERRAGOSTO PARTY

THIRTEEN

Ferragosto

IF FERRAGOSTO PUTS THE REST OF ITALY INTO SLOW MOTION, in Sicily it induces total paralysis. Bernie, Damian, and Tania have departed, and Signor Cassano has bidden his farewell a week in advance to an undeclared destination. He has told me pointedly that he will be leaving his mobile phone behind. This is unfortunate, as on the very day of his departure our brand-new pump breaks down, leaving us high and dry at the height of the summer without a drop of water. Meanwhile the Husky plumber has vanished into thin air. He has never been an easy catch. On one occasion in an attempt to track him down we even resorted to lying in wait outside his beach apartment, but to no avail. (He eluded us, but we did manage to talk to his teenage daughter, who apologized for the evasiveness of her "unpredictable" father.) As for contacting him on his holiday, there is no point in even trying. We will have to seek help elsewhere. Trying to get hold of a plumber on a Friday night is a challenge at the best of times, but on the run-up to Ferragosto in Sicily, it simply doesn't bear thinking about. But, not to worry, Marcello is confident that the Family can sort out the pump problem. After all, on site at Santa Maria we have an ingegnere—Pilli—an architetto—Silvio—a geologo—Goffredino—and a student *biologo*—Emiliano. All four are summoned for consultation. They stand over the pump with its conglomeration of tubes and switches and discuss the issue loudly. Emiliano is the only one to get down on his hands and knees. He fiddles about with the controls, but the Husky's state-of-the-art automatic pump proves to be too much of a challenge, and he has to admit defeat. The architetto and the ingegnere shake their heads in a gesture of Sicilian resignation.

"*Niente da fare.* There's nothing you can do until after Ferragosto." Generously, Silvio reminds us we can use the hosepipe in the cortile to take care of personal hygiene.

"Well it's better than nothing!" As an afterthought. "If necessary, you can use our bathrooms."

Marcello can barely disguise his annoyance. As for Clem, hell has no fury to match that of a teenage girl denied use of the bathroom. Hair-washing plans for the peak of the summer party season are seriously threatened, thanks to gross parental failure to maintain vital hydraulic equipment in working order. The combination of no water, temperatures hitting 100 degrees Farenheit, Marcello in Manzo Temper mode, and teenage daughter on the rampage is not a happy one. Fortunately, Lul detonates the situation before there is a major explosion by ferreting out a plumber, possibly the only local plumber who is willing to flout the Ferragosto boycptt on gainful activity. The pump is functioning again within twenty-four hours. Relieved, showered, thoroughly revived, and grateful to Lul, the plumber, and the Virgin Mary (Ferragosto being her feast day), we can look forward to the festivities ahead.

What festivities and where they will take place has not yet been discussed when Lallo and Rosetta announce a decision to Get Close after lunch. Horrified to find us lunching in the garden—Are you crazy?—they persuade us to retire to the cool of the dining room for coffee. Sensible Lallo and Even More Sensible Rosetta, although by no means the most senior of the cugini, are the unofficial Elders of the Family, thanks to their consistent *buon senso* and infinite supplies of good advice. Not that we have solicited any advice to date, our main preoccupation being what to cook for dinner each night, and on this subject we have Pilli, our resident gourmet for on-site consultation. No, the Sensible couple has come of their own accord. Is this just a friendly visit, or do they have something of importance to say? It turns out that they do.

As they stoically down my offering of filtered coffee—*ugh! caffè americano!*—the conversation comes round to the topic of Ferragosto. Never having been in Sicily before at this time of year, I hadn't realized it was an issue at all until now. As Lallo explains, the Vaccara Family and Honorary Family such as Aldo and Lelle, Rino and Vera, always celebrate the night of the fourteenth together.

"What, you mean a party?"

"*Sì, una festa.* We usually take turns to host it."

"Whose turn is it this year?" There is an awkward silence.

"Weren't you going to have a *festa di inaugurazione* anyway?" The penny finally drops. Everyone has been waiting for Marcello and Caroline to invite them to a grand housewarming on the eve of Ferragosto, a double whammy of a celebration.

"How many people are we talking about?" Lallo makes a quick calculation.

"*Una cinquantina.*" Fifty people to dinner at the hottest time of the year and no one to help out in the kitchen. Even Marcello looks daunted. He reminds me that Susan and Manlïo, our friends from Nairobi, are arriving on the thirteenth. By this time a delegation has arrived from Next Door and another from Upstairs. Has there been an element of planning in this casual Getting Close? Maria Antonietta senses our predicament.

"You won't have to do all the cooking yourselves. Pilli and I will help you. We'll do the pasta—maybe pasta with ragù." Lallo chimes in.

"And Aldo and I will help Marcello with the barbecue."

"And of course every family will bring a *dolce*," adds Rosetta.

With the menu decided and responsibility nicely distributed over the collective shoulders of Family, I am suddenly looking forward to hosting our first festa. So are Marcello and Silvio.

"Do you realize it will be the first festa in the cortile for over twenty years, since Papà (Goffredo) died, in fact?"

"*Bravi. Buona decisione,*" says Rosetta approvingly. Then, in case we change our minds, "Let's make out the guest list now." In no time the impromptu committee has come up with a guest list, consisting mainly of a hard core of Vaccaras and Honorary Family, with some Manzo cousins thrown in and a handful of transient guests. Silvio would like to invite his shooting and bird-watching cronies. Nanette proposes a few names from the Sicilian American community, but Lallo and Rosetta wield the whip. Unless we limit this to Family, the number will escalate out of control. Feeling a bit superfluous to what, after all, was also to be our housewarming party, I make a suggestion of my own.

"Why don't we make it *in costume?*" There is a stunned silence, followed by an animated debate culminating in a majority vote in favor. The Flowers, nostalgic for the Rio Carnival, are beside themselves with joy at the prospect of a fancy dress party. My reward is Lallo's blessing. Before he leaves, the Family Elder stoops from his lofty height to put an arm around my shoulder.

"*Brava Caroline. Una buona idea.*"

With Ferragosto, the climax of summer festivity, just days away, it strikes me that this is somewhat short notice to get the whole Family together. Despite the late notice, not one of the invitees turns us down. Some of them even sound surprised that we are calling them at all.

"Of course we're coming." It sounds as though there has been forewarning of this last-minute invitation, and Santa Maria is where they were expecting to celebrate Ferragosto all along.

"*Ormai ferragosto si fa a Santa Maria*—Santa Maria is where we'll be celebrating Ferragosto from now on." The last twenty years have just been a mere hiccup, and now that the ground floor has been renovated and all three brothers have made a home at Santa Maria, the tradition of parties in the cortile, established by Maria and Goffredo, will once more be resumed.

With a mammoth catering task ahead of us, clearly there has to be a division of labors. Marcello immediately puts himself in charge of supplies, the ordering of the sausage, the beef for the ragù, and enormous quantities of handmade *pasta busiata*, which comes in the shape of double strands twisted like corkscrews. The organization of the tables is delegated to Nanette, the Flowers, and me. I offer to procure the necessary tableware and cutlery, borrowing from the Luigi wives if necessary. Nanette stares at me in disbelief,

"Nobody expects us to provide regular plates and glasses. Just get plastic."

"What, even the knives and forks?"

"Plastic. That way nothing gets broken and we don't waste time washing dishes."

"Anyway," Stefi pipes up, "Plastic is a party tradition. People would think we were stuck up if we put out posh plates and stuff." Tradition or not, I agree that Sicilian priorities have been applied here: food first, appearances second.

"What about the wine?" I seem to be the only person concerned about this, and my bring-a-bottle suggestion does not go down very well. Asking people to bring wine would most definitely create a *brutta figura*—bad impression. Sicilians are far more interested in eating than drinking, so we decide to furnish a few choice bottles backed up with supplies from the Cantina Sociale, the wine outlet in Via Maccagnone attached to the ex-Vaccara factory. Now it supplies the bottom end of the local market with rough Grillo and Cabernet from large metal contraptions, which look suspiciously like old-fashioned petrol pumps.

With guest list, equipment, food, and drink taken care of, the Flowers and I address the problem of what we are going to wear. A trunk of party clothes, Brazilian carnival costumes, and Maria's donations of '50s and '60s finery are hauled out of the warehouse. In my new role as wardrobe mistress, I distribute the contents. Melissa, Clem, and Stefi are well provided for, but there is nothing for Massimo but a Pierrot costume that Marcello once wore to a fancy dress party in Jakarta. Turning up his nose at satin and pom-poms, Massimo decides to go for something more manly. He will be a Greek hoplite, and he will make his own outfit. As for Marcello, he is too engrossed in preparing for the party to give much thought to dressing up for it, but on a visit to a kitchen shop in search of pans, his eye is caught by a professional chef's uniform—white coat and apron, checked trousers, and chef's hat. Just the thing, and, since he will be chief barbecue minder, very appropriate too. He decides to go as a chef.

"Mummy, what are *you* going to wear?" Clem is worried that her mother has not been decked out. By now I have seen Stefi and the Flowers emerge as Cinderellas ready for the ball in the glad rags of my yesteryear. Certain to be upstaged and loath to appear as the Ugly Sister, I have decided to go as a Sicilian widow. But where to get hold of Sicilian widow's weeds? To borrow from Maria for a fancy dress party would be in bad taste. Lelle has the answer. Mazara abounds in shops that come under the category of *casalinghi*—houseware stores. Perhaps one of these will stock what I'm looking for. Sure enough, the first one I try has its own Sicilian widow's weeds department—rails of black dresses, shawls and scarves in all shapes and sizes. Don't forget the gold bracelets, says Lelle. And for an authentic touch, I will pencil in a fuzzy mustache.

Meanwhile, our friends Susan and Manlio have arrived. Forewarned, they have come forearmed, Susan with blue nylon wig and Manlio with false nose, mustache, and bowler hat. Whatever they are supposed to be nobody is quite sure, but they have clearly entered into the spirit of the thing. On the day of the party Silvio appears downstairs with the components of a complicated costume that requires the help of a seamstress. Would the Flowers or I be prepared to do a bit of sewing? What with all the activity going on in the kitchen and tables still to set up, his request is inopportune to say the least, but fortunately Susan comes to the rescue and sits down to sew on the various buttons, frills, and ruffles that go with the outfit. At the last minute Nanette comes down, flustered from the flurry of Ferragosto preparation. She hasn't organized her own costume yet. The trunk by now is almost empty, and she has to make do with a kimono and a conical peasant's hat from Sulawesi, which sits oddly out of place on her thick blonde tresses.

The tables are set up, the barbecue burning nicely, and the court-yard glowing with the light of lanterns rigged up by Emiliano. The first guest to arrive is Joe, Nanette's father, a sprightly ninety-year-old in his normal gear and a banded hat.

"Nonno, who are you?" asks Stefi.

"Un mafioso." He looks the part without having had to dress up. I come to greet him.

"Baciamo le mani." I bend to kiss his hand, the Sicilian widow acknowledging the power of the local henchman. More guests arrive. Luigi Emilio, another mafioso, is carrying a violin case. Enza, his wife, plump and sixtyish, decked out in green silk bloomers, skittishly thrusts a gold-painted face at me.

"Chi sono? Chi sono?" I have no idea who or what she is supposed to be. A pumpkin? A green tomato? A peacock?

"No, no, stupida. Sono la Primavera."

Simonetta, their daughter, wears a minimalist version of Indian squaw gear, displaying long, beautifully toned legs, guaranteed to pro-voke reactions ranging from desire to death glares. Lelle, her sister-in-law, Maria Teresa, and Rosetta, in seductive veils and chiffon, form a mini-harem. Massimo appears covered in armor of overlapping sheets of aluminium foil, looking less like a Greek warrior than the Tin Man from *The Wizard of Oz*. The barbecue is manned by Pilli in the guise of Asterix, the Gaul, his paunch protruding below a cardboard breastplate, and Aldo, *il Capitano,* wearing nothing but a giant diaper, incongru-ously impersonating a *bambino.*

While his brothers sweat over the barbecue, Silvio, transformed into a milord of the ancien régime, in breeches, white stockings, and tri-corned hat, poses for photographs with the Flowers and Stefi, all three masked, gorgeously arrayed in hot pinks and reds and crowned with feathered headdresses, as if parading for a 1930s Rio Carnival. But the star of the show is Gianluigi, son of Luigi, the Baron. Tanned and green eyed, with bunches of grapes hanging from his curly blonde locks, this sultry Dionysus circles us, menacingly tapping his thyrsus on the stones of the cortile, while Lul and Michela, his attendant maenads festooned in vine leaves, perform a frenzied dance around him.

The pasta is cooking in our kitchen and the barbecue is heating up nicely when suddenly the hum of conversation is interrupted by a thudding drumbeat. The *portone* is flung open, and in bursts a seraglio of veiled houris led by Maria Antonietta, rolling her eyes and writhing with bared torso in an Egyptian belly dance. Eros beating a drum, and Emiliano and Goffredino leaping and whirling like dervishes, follow their mother as she circles the courtyard seductively swaying her hips and spreading her veil in invitation to onlookers to join the frenetic rhythm of the dance. Emiliano—lean, long-limbed, and clad in black lycra—suddenly leaves the dance to scale the walls over the arches, hauls himself up onto Silvio's terrace and, dervish turned Spiderman, has the rest of us open-mouthed as he swings from one of the towers.

At midnight the Young Ones depart to converge on the beaches around Granitola and Torretta, where they will light bonfires and see the dawn in to music and dancing. Three hours later we say good-bye to the last guests, ushering them through the portone and acknowledging their effusive thanks. For Enza, Luigi Emilio, Simonetta, and her fidanzato, the festa has been a success.

"*Ormai ferragosto si fa a Santa Maria*," Enza reminds me, looking forward to future celebrations. Luigi raises his mafioso hat.

"*Bravi, bravi!* It was just like Santa Maria in the old days."

Something is on Simonetta's mind. Before she gets into the car she draws me aside and whispers in my ear.

"*Sai, Caroline*," discreetly pointing to my shaded upper lip,

"There's no need to put up with facial hair—you can do something about it. Now there's one very good product I can recommend . . ."

"FAI LA SALSA QUEST ANNO, CAROLINE?" MARIA TERESA IS ASK-
ing me if I intend to participate, not in a sexy Caribbean dance, but in
a staid southern Italian tradition—the annual tomato sauce bottling ritual. We are at dinner at Aldo and Lelle's house at Torretta. The men are
settling into one of their *dopo cena* political discussions-cum-shouting-
matches, while the *signore* home into more pressing issues—in this case
la salsa. Aldo's sister-in-law proudly announces she has made 150 bottles
with only her son to help her. Why so many, I ask. Are she and her husband going to be eating tomato sauce every second day for the rest of
the year? She looks at me as if to ask which planet I am coming from.

"*E i figli?*" Silly me! I hadn't even given them a thought. Their children, who are at university in the north, will be driving back up there
at the end of the month with a hundred bottles of tomato sauce, fifty
each to last them through the year. That way Mamma and Papà can be
sure that the staple diet is provided. And there will still be fifty bottles
left over for the parents. Pangs of guilt start threatening to assault me,
but the thought of the Flowers setting off for London lugging fifty bottles of tomato sauce apiece is so preposterous that it fends off remorse
before it sets in for real. Nevertheless I have to bear the shame of
neglecting the welfare of our two daughters, both of whom will be living on a shoestring in Spartan student conditions without a single bottle of homemade tomato sauce to fall back on.

"What about your girls? How will they survive?" The poor things
will be forced to buy tomatoes out of a tin. I might as well admit it; I
will never make a true Sicilian mamma.

The peak season of the pizzutellii, dark red, thick-skinned Sicilian
cherry tomatoes, the best ones for salsa, coincides with the hottest week
of the year, the third week of August, but despite the heat, all over the
Sicilian countryside families are doing what their parents and forefa-
thers have always done year after year—laying up provisions for the
coming months. I have been vaguely aware of this annual event, but, as

a visitor, I have never been invited to take part. Our new status as homeowners now requires us to join the tomato-bottling club. In view of our ignorance, Maria Antonietta decides to take us under her wing and invites us to join Next Door's salsa-making on an apprentice-ship basis. All is ready for a mammoth salsa session with a hundred empty beer bottles, washed and sterilized ready to be filled with the red-hot liquid. But what about the tomatoes? Usually she buys from our neighbor, who has a vegetable patch alongside his vineyard, but this year she is too late. He has already sold them to another customer, which is a great pity, because the terra of Santa Maria produces the best tomatoes anywhere. Having scouted the local markets, she has found a signora in nearby Campobello whose tomatoes meet the non-watery standard she is looking for. Five crates should be enough for a hundred bottles.

"Come over tomorrow and help us. Then you'll be able to do it yourselves. After you've made your own sauce, you'll never want to buy tinned tomatoes again." And of course salsa made out of season with greenhouse tomatoes is completely beyond the pale.

In view of the quantities involved and the size of the apparatus, the rite has to be performed outdoors on Pilli and Maria Antonietta's ter-race. A few years ago I would have been horrified at the thought of spending a whole day of my holiday sweating over simmering tomatoes under the August sun. Now I am looking forward to it as an important stage in my Sicilian education. Two great cauldrons are placed over gas burners, one for cooking the sauce, the other for sterilizing the filled bottles. Work starts with Emiliano in charge. Into the cauldron go chopped red onions, fistfuls of basil, and lots and lots of tomatoes. Once the sauce is pronounced ready, he drains the contents of the pot with a slotted spoon. It's impossible to strain the tomatoes in the normal way, as no colander would be big enough. Next the cooked tomato and onion mixture is passed through the mill to obtain a puree. Pilli turns the handle, grunting with the effort, while Marcello decants the thick red liquid through a funnel into the lines of waiting bottles. While Pilli and Maria Antonietta argue over every detail at the tops of their voices, Marcello plays the buffoon, pretending to burn himself and drop the bottles. Everyone plays an important role in the process, and not wishing to appear completely useless, I find work—washing tomatoes, chopping onions, and sweeping up the mess on the terrace. It is Emiliano, however, who keeps the whole operation under control. Standing quietly in his swimming trunks over the simmering cauldrons, he checks the cooking time, decides just when enough water has been evaporated, and makes sure everything is in place for the next stage of the process.

The sauce-making goes on for two days, with the whole family taking turns to stir, strain, and decant. Even Goffredino, who has only been able to snatch a few days of holiday from his new job, is happy to join in. Lallo and Rosetta drop by and end up giving a hand, although they have already sweated over their own year's supply of salsa. As an apprentice, my part in the process has been minimal. Do I see myself performing this task every year for the rest of my life? In theory yes. Homemade tomato sauce is a Very Good Thing. But can I, or rather, Marcello and I, do it alone?

Having undergone two whole days of full immersion in salsa making, our determination is immediately put to the test. We will make our own salsa, albeit on a far more modest scale than that of Next Door, with just two crates of tomatoes. Maria Antonietta has promised to lend us her equipment and supervise. The rest is up to us. Naturally, freshly harvested *pizzutelli* do not come in ready-for-salsa mode. Leafy stems must be removed and the tomatoes washed. This is not as easy as it might sound. The crates, weighing forty pounds each, cannot easily be lifted so, Marcello not being available because of more pressing duties, I decide to leave them on the ground. This means bending down to pick the tomatoes up one by one, remove the stalks, and chuck them into a bin ready for washing. This backbreaking start to the operation takes roughly one and a half hours, by which time my hands are pitch-black with tomato dirt. Marcello returns from his early morning mission to town (having satisfactorily met his objectives of buying bread and newspaper and having a coffee with a *compagno di scuola*), just as this task has been completed. I leave him to hose down the bin of tomatoes while I retire to the kitchen to wash a bucketful of basil and chop ten pounds of onions. Taking a tip from Melissa, who bursts into tears at the very sight of an onion, I don builder's goggles. Maria Antonietta appears at the fence, ready to pass the equipment over. She is horrified to hear we don't have a gas cylinder handy. This means we can't use her gas burner. We'll have to cook the sauce on our stovetop, which will take twice as long. Over the fence she passes two huge tin pans, each big enough to bath a baby in. Then comes the *mestolo*—a three-foot-long wooden spoon, a tin ladle, and an outsized food mill for separating the sauce from the skins. All items being of giant proportions, they look like stage props for a performance of *Jack and the Beanstalk*. The only normal-sized piece of equipment is the bottle-capping machine.

Maria Antonietta, the maestro, arrives in our kitchen ready to start us off. She fills the two pans with tomatoes, the chopped onion and basil and several generous handfuls of salt. Quantities are not measured. I'd never dare to ask her for them. She's been making salsa ever since she was knee-high and knows by instinct just how much to put in. At full flame the cauldrons are left to their own devices for ten minutes or so until they start to *fare acqua*—make water. At this point, Maria Antonietta grabs the Jack-and-the-Beanstalk spoon and immerses it into the seething mass of red, making sure it scrapes the bottom of the pan. Then I am allowed to take over, stirring at regular intervals as instructed, so that the tomatoes don't stick. I wonder whether I can do a bit of multitasking between stirs, make a phone call or clean the bathroom perhaps. Or do the steaming pans require my undivided attention? No, my sister-in-law makes it quite clear that I should not take any risks. It's simply a question of priorities. Once the simmering is well advanced, I embark on the *schiacciamento*, the squashing of any tomatoes stubborn enough to resist the gas flame at full blast. For an hour I keep an eye on the bubbling red lava, stirring and squishing by turns. Marcello now takes over. This operation is much more to his taste than the cleaning one that preceded it. I decide it doesn't take two to stir the tomatoes, so I retire temporarily, but not before taking a picture of him stripped to the waist, in nothing but swimming trunks and his chef's apron, stirring away with both hands gripping the mestolo, like a mustachioed Circe at her cauldron.

At midday Maria Antonietta reappears. She declares the salsa to be of just the right consistency after an hour and a half of simmering. The cauldrons are carried to the tables under the orange trees for the straining and bottling process. Marcello wrestles with the mill, which he has attempted to screw on to one of the trestle tables. However hard he turns the screws, it wobbles in an alarming manner, so he calls Goffredino, an old hand at the tomato-milling process. Goffredino climbs over the fence in a sarong (and no underpants, he warns us). Marcello fools around with the ladder so Goffredino has to lift his skirt. Much avuncular laughter ensues. When nephew discovers how his *Zio* has set up the mouli contraption, it's his turn to laugh. With a noisy guffaw he chides him for attaching it back to front. After a hearty uncle and nephew slanging match, the mill is attached, complete with screw and filter dish. We place a pan on a chair to catch the freshly milled salsa, and we're ready to go.

Goffredino ladles the tomatoes from the cauldron into the mill. Marcello grinds away so that the skins, onion, and basil are spewed out onto the filter dish, while the thick red sauce oozes down onto a ladle held by Maria Antonietta, and into the pan. An amazing amount of watery residue is left in the cauldron, which we throw into our garden to nourish the citrus trees. Ready at the next table with the funnel, I fill each bottle with the hot red sauce and pass them to Emiliano, ready with the bottle capper. In twenty minutes we have milled, bottled, and capped it all. But this is only the first round. There are another forty pounds of tomatoes out in the courtyard waiting to be converted into salsa.

The cooking operation has to be repeated for the second batch. Maria Antonietta, Goffredino, and Emiliano go off for lunch, the two

boys crossing the fence by ladder, their mother taking the more digni-
fied route through the parco. Marcello and I are experienced enough
now to tackle the second batch on our own. Marcello does his Circe
act again with the gigantic wooden spoon. Just watching him is enough
to make me wilt. By this stage I look and feel as hot and red as the
tomatoes in the cauldron. It's now two in the afternoon, with the temperature in the kitchen several degrees higher than outside, and here we
are, standing over forty pounds of bubbling tomatoes. Any sensible non-
Sicilian Italian, commenting on an apparently pointless feat, would say
"*Chi ti lo fa fare?* Who's making you do it?"

By three o'clock the household next door have had lunch and
returned for the second bottling operation. By four it is all over, and
we proudly count twenty-five bottles of nice thick salsa, a good yield
for the quantity of tomatoes. My sister-in-law reminds us we should
add sugar when we open the bottles and reheat for at least half an
hour. Garlic and fresh basil may be added at this stage. But our toils
are not over until we have sterilized the bottles. First we wash all the
equipment and put the two cauldrons back on the stove. Then on
Maria Antonietta's instructions we wrap the hot bottles in dishcloths
and pack them in to the great pans. Of course I don't have twenty-
five dish cloths handy and have to borrow hers. The water has to boil
first, which takes ages with all those rag-wrapped bottles bobbing
away and the bottles must simmer for twenty minutes; after which
they are left to cool for twelve hours before they can be removed and
stored. We have toiled and sweated for the best part of a day, but it was
worth it.

As we are proudly contemplating our twenty-five green bottles,
Diana appears at the fence. Unlike her sister, she takes little interest in
food preparation and has never made salsa before. When she comes to
Mazara, she prefers to go to the beach and see friends rather than sweat
over a hot stove. Having watched us beginners with interest, this
evening she announces her attention to make some salsa herself.
Whether Luca is willing or not, she's going to give up the next day at

the beach to make her own tomato sauce. "Call it a *raptus*—irrational fit of madness—if you like, but there's no excuse for me not to do it now. After all, if it can be done by *un'inglese . . .*"

ACCORDING TO MARIA'S DIVISION OF THE PROPERTY, THE *parco*—the garden in front of the villa—is jointly owned by Marcello and Silvio. Marcello, the absentee landlord, is ever critical of what he sees as negligence on the part of his younger brother, co-landlord and resident custodian. One evening as we are walking the dogs, the sight of a generation of accumulated neglect becomes too much for Marcello to bear. He deals with his frustration by vigorously cursing the weeds and the litter and snapping off dead branches from desiccated pine trees, pausing now and then to make glum calculations of just how much of the Nest Egg a clean-up operation would devour.

"Call this a parco? More like a jungle. Why can't my fool-head of a brother *do something* about this?

"Well, he has quite a lot to do, you know. Apart from his job he does all sorts of stuff—painting, bird-watching, shooting, bookbinding, stamp collecting—"

"Bookbinding, stamp collecting! What on earth is the use of that? He lives here, doesn't he? Why doesn't he do something useful? What about a bit of gardening *per amor di dio?*"

I have to admit that Silvio, extremely knowledgeable on the principles of gardening, has rarely been seen putting those principles into practice. As for Nanette, she has many domestic interests, but gardening is not one of them.

"Look Marcello, this isn't getting us anywhere. If Silvio's not interested in cleaning up the parco, why don't we just go ahead and do it ourselves?"

"How on earth can we, if we don't even live here?" Short of spending a lot of money to get someone else to do it, we can't. In a despondent mood we make our way back to the portone, where we find Silvio,

Nanette, and Massimo unloading trays of geraniums from the back of their car.

"What's all this?" Why the sudden plant-buying spree?" Massimo cannot contain his excitement.

"It's for the Bishop! He's coming to visit us!" The flowers are for decorating the courtyard where Silvio and Nanette will entertain the *vescovo*.

Seizing the moment, I innocently interject, "Hmm, it's all very well tarting up the cortile. But what about the front of the villa? The Bishop's going to see all the mess as he drives in. We don't want to make a *brutta figura*, do we?"

"Gee, Silvio. Caroline's right. At least we could clean up the driveway, couldn't we?"

Nanette and I have now sowed a seed of doubt in Silvio's mind. He has cultivated his rapport with the Bishop of Mazara for some time, and now that Sua Eccellenza has accepted the invitation to visit, he too sees that the entrance to Santa Maria, choked with weeds and debris, might very well make a brutta figura. Marcello and Silvio make a last-minute attempt to hire a gardener, but they have left it too late. At this point there's only one solution to the garden problem. We'll have to clean it up ourselves after all.

The following day is Saturday, and we decide to do a seven to ten stint so that we can lay down tools before it gets impossibly hot. The equipment is set out in the entrance porch the night before: hoe, saw, rakes, spades, shears, wheelbarrow, and a pile of gardening gloves. And the strategy? I insist that we should have one. Sicilians dislike anything that smacks of organization, but, I argue, it's no good if everyone potters around doing his or her own thing. After long debate, in the end we agree to a plan: Marcello and Silvio will rake the front garden (not all of it, only the part visible from the drive), remove the fallen palm branches, heap up the debris, and carry it to the bonfire, of which they will be in charge, bonfire management like barbecue-minding and knife-wielding being an exclusively male prerogative. I suggest that

Melissa supervise her younger cousins shaking out the bougainvillea and sweeping up the dead flowers and leaves in the drive. Clem refuses to commit herself as to her intended role. As for Nanette and me, we will tackle the weeds—uproot them with the hoe and sweep them into piles ready for a bonfire.

I set our alarm for 6:30. Marcello and I are the first out into the garden at seven o'clock sharp. Melissa follows us bleary eyed.

"I feel sick."

"Well go back to bed then."

"No no." and she stumbles past, tripping over the wheelbarrow. Minutes later Silvio appears in what looks like fancy dress—ye olde farm laborer in green dungarees, battered straw hat, and gardening boots. Nanette arrives wielding a mighty hoe, but somehow in a little black dress and sandals she looks wrong for her intended role. Marcello, bare chested in his swimming trunks is spoiling for the fight, as though he had been reading Hesiod, who advised his readers to "strip to plough and strip to sow." No doubt the Greek poet knew a thing or two about gardening in the heat of the Mediterranean summer.

Naturally the whole party has had a late night, all nights being late in a Sicilian summer, so no one feels particularly chatty. We slog away at our assigned tasks, but the teamwork between me and Nanette is hampered by the fact that there's only one hoe—the one Nanette is using. Now literally into the swing of things, Nanette in her cocktail dress is wielding the hoe like a mad dervish. There are many expletives of *Matri mia!*—Mother of mine!—the nearest my sister-in-law will ever get to swearing, as she stubs a toe on a rock or succumbs to a stab of back pain. Without a hoe of my own to swing around, I get on my knees to tackle the weeds, going for the roots with a trowel. Marcello has already been at them with weed killer, so the task is easier than usual. Vast clumps of summer weeds—canes and wild grass mostly—come up at first tug, but blackberry brambles, which have somehow rooted themselves in the road, put up fierce resistance. This is a battle we are determined to win. While my sister-in-law heaves and hacks at the *erbaccia* like an amateur executioner, I crawl along behind her on all fours, fiercely digging out any recalcitrant roots that have evaded her frantic efforts.

It's quite a bucolic scene really, with Nanette swinging her hoe, me apparently communing with nature at ground level, Stefi shaking bougainvillea branches, Melissa sweeping, and Massimo toing and froing with the wheelbarrow. But where have Silvio and Marcello gone to? If they can't be seen, they can certainly be heard engaging in what sounds like argument but is probably just fraternal banter. The peaceful scenario is interrupted by Chris, Silvio's springer spaniel, who cavorts in circles around us, batteries fully charged, and Hans, Nanette's Doberman annoyingly tearing up and down the driveway, pausing only to sniff and slobber. At one point he actually leapfrogs over my crouching figure. I dare not voice my opinion that we'd be better off without the doggy participation.

Maria Antonietta strolls over from Next Door in a décolleté dressing gown and mules, smoking a cigarette. What are you doing? Who's making you do this? She can't think of a worse way to spend a Saturday morning. Look at Caro*line* on her knees, she shouts to Luca as he arrives en route for the beach, leaning out of the car to display a lean, tanned torso. They gape and marvel for a bit, then disappear. Minutes later they're back, ready to join the team. Maria Antonietta dressed as the Glamorous Gardener in snow-white stretch T-shirt, pedal pushers, and gold sandals, grabs a rake while Luca takes charge of the wheelbarrow. Neither of them is the type to work in silence, so Maria Antonietta attaches herself to Nanette for girlie gossip, which slows down our coordinated weed extraction activity considerably, and Luca heads off in Silvio's direction. Minutes later Luca and Silvio have put down their tools and are sitting on the garden wall laughing and cracking jokes that have nothing to do with gardening.

Three hours later we have finished. A huge blister has appeared on my right palm and the skin has come off both knees. Nanette is speechless, unable to pronounce anything but an agonized *Matri mia*. Melissa, red-faced and frazzled, can barely stand up, while the children look as shaken as the bougainvillea they have been beating with sticks. Silvio and Luca, having abandoned any pretense of active participation long ago, are contemplating the results of our efforts and drawing up plans for the next stage of a Clean Up Santa Maria project. Maria Antonietta, summoned by Goffredino, has gone to make breakfast for the men folk Next Door. The only person who has not ground to a complete halt is Marcello, who keeps going with the saw, hacking away at dead pine branches. I know my husband. Once he gets started he doesn't know when to stop.

"Come on Marcello, that's enough."

"I'm coming *ora, ora*," he says, *ora* being Italian for "now," but Sicilian for "in my own time." As we shuffle through the portone, filthy, sweat-soaked, and stinking to high heaven, a lovely vision appears before us—Clem, deeply tanned, in bikini top and minimalist shorts, freshly bathed and scented.

"Mum, *Papà*. Don't tell me you've finished already. Why didn't you *wait* for me?"

In the end our exertions have rendered the driveway and front garden fit for a bishop's scrutiny. The danger of making a brutta figura has been averted, and the visit goes smoothly.

"Just as well he didn't ask for a tour of the grounds," Stefi whispers sotto voce. I agree. It would have been highly embarrassing to reveal to our visitor the anarchy that prevails in the gardens surrounding the villa. As it is, His Worship pays Silvio and Nanette "*tanti complimenti per il castello* and the *parco bellissimo*."

SANTA MARIA FROM THE ORANGE GROVE

FOURTEEN

Christmas in Sicily

IF WE EVER THOUGHT THAT BY RETIRING TO A REMOTE COR-
ner of Sicily we would enjoy a quiet Christmas and New Year, we were
hopelessly wrong. By early October the Milan-Mazara-Cairo hotline
has been set up by my dear sisters-in-law, who are plotting a major celebration at Santa Maria. Celebration? What? News has leaked through
that Pilli's latest business venture has floundered and his wing of Santa
Maria and adjoining grounds have been reclaimed by the banks. With
their home about to be confiscated, how come the Pilli Manzos want to
celebrate? And what about Silvio and Nanette? Are they not beset with
worry that the future of Santa Maria might be at stake? Well it's not as
bad as that, Marcello points out. Given the grinding slowness of the Italian legal system, it will be months, possibly years before Pilli's assets go to
auction. Besides, he has just signed a lucrative contract to work in the
Far East. In a couple of years he will have made enough money to buy
his house back. So don't worry. Let's go ahead with the party.

The festa committee has decided on a buffet spread upstairs, then
fireworks and champagne at midnight in the cortile, followed by dancing

Next Door—all in evening dress. Why, I groan. Didn't we do our bit at Ferragosto? Do we really need to have another big bash at Santa Maria? (And don't we have other worries on our minds? I lack the courage to add.) But I have overlooked some very good reasons for the party: it will be a celebration of Goffredino's engagement and Maria Antonietta and Diana have already bought long dresses for the occasion. Maria Antonietta on a crackling line from Cairo does her best to kindle some party spirit.

"You'll see, Caro*line*. We'll make it so elegant. It'll be like the *ballo* in the *Gattopardo*."

For the Marcello Manzos there is a touch of sadness this season, as this will be our first Christmas without Clem. She has accepted an invitation that no eighteen-year-old could refuse—to see in the New Year in Rio de Janeiro. In recompense, Melissa flies back from Tokyo and she joins me and the dogs at Milan airport to catch a flight to Palermo, while Marcello stays behind in Milan for "end of year duties"—i.e., the company Christmas dinner.

This year Sicily has seen no rain for nine long months but, as luck will have it, the heavens open just as our flight lands at Palermo airport. We are met by a local taxi service, and no sooner have we hoisted the dogs in their cages into the back of the van than we realize we are in for a storm. At first it's just heavy rain, not so much drops as pellets, which then upgrade to hail the size of rocks, battering away at the windows like a demented drunk locked out from a party. By the time we are on the motorway, visibility is almost nil, but the speedometer is hovering at an alarming 90 miles per hour, well over the official speed limit. Panic sets in when I can't find my seat belt. I try to locate it as discreetly as possible, passenger use of seat belts being considered a vote of no confidence in drivers over here. While I carry out a furtive hand search, our loquacious driver treats us to the story of his life. He tells me he was an assistant to the Italian consul in Saint Louis, Missouri, for many years. This is puzzling, since it is quite clear that he doesn't speak English.

"Assistant?" Staring at him as I grope in search of a seat belt, "Was it an official position?"

"Not exactly official but I was his interpreter. He paid me good money."

"Interpreter? Of what?"

"*Siciliano!* Most of the Italians in Missouri in those days were Sicilians. They spoke dialect, not Italian of course and the consul was a *polentone*—a polenta eater." By polentone he means a northern Italian who doesn't speak Sicilian. I still can't find the seat belt and I'm not comfortable with the driver's habit of using his hands for emphasis. Sicilians being more emphatic than other Italians, the driver removes his hands from the wheel farther and for longer. Turning his attention to Melissa, hunched into a corner under an overhanging backpack, he tries to entice her to take part in the conversation.

"*Che c'è, Signorina*—What's the matter? If you don't speak, everyone will think we've had a row." Swiveling right round at this point to see if he's raised a smile, both hands come off the wheel. As we belt along at 100 miles an hour in the blinding rain, I thank God, Santa Rosalia or whoever is the patron saint of passengers, that there is no other traffic on the road. But if I can't get the driver to shut up and slow down, I must at least find the seat belt. Realizing that I risk offending his pride in his driving prowess, I blurt out my problem. He couldn't be more helpful. With just his left hand on the wheel, he stretches his right arm in my direction, locates the hidden seat belt from somewhere behind my head, pulls it down and fastens it himself.

"*Voilà!*"

The banter and energetic body language continue until we reach the end of the motorway. Melissa manages to sleep or at least to feign it. The storm shows no signs of abating and our arrival in Mazara is hailed with rolls of thunder followed by dramatic bursts of lightning. This has an unexpected effect on our cheery companion. Miraculously he slows down. Then he starts talking, or rather, moaning to himself, "*Mamma mia! Matri mia! Matruzza mia! O la Madonna!*" the pitch of his voice rising with each permutation on the Mother theme. A clap of thunder has him sinking down into the seat; at a particularly violent crack of lightning he starts to shake. The poor man is clearly terrified, and ashamed of the fact.

"*Scusi, Signora*, I've been afraid of lightning ever since I was a *piciriddru.*"

We are off the motorway now so I tell him to park on the side of the road. He does so and we sit out the storm together in the pitch dark. While Melissa sleeps, I try to calm down our driver with all the skills at my disposal—take deep breaths—focus on something positive—your wife, your family. Aren't they expecting you? It's no good. He starts up a low wail, which rises to a shriek every time a white flash in the sky lights up the warehouses of the local pasta factory.

Silvio, Nanette, and their children are waiting at Santa Maria to welcome us. Having said good-bye to our driver and made a mental note of his name to avoid booking him in future, we let out the dogs, dump our luggage, and head upstairs for supper. Like many homes in Sicily the upper floor of Santa Maria does not have central heating. In spite of the gas heaters the sitting room is decidedly cold, so Maria has been moved to a smaller room, which is easier to heat. Her cheeks pink from the warmth, she seems frailer than ever in winter, hunched inside layers of fleece and wrapped in a cashmere shawl. Aware of the presence of newcomers, she smiles up at us and returns our embraces but there is no flicker of recognition in her eyes, now faded to a milky gray.

Marcello's plans to install heating on the ground floor before we arrive have gone awry, so we have no source of warmth on our first winter night except an electric heater courtesy of Nanette. After an uncomfortable night, made just bearable by sharing a bed and the only duvet in the house, Melissa and I wake to find it is still raining outside. The reality of a Sicilian winter in a house without heating now sets in. The large rooms with their high ceilings and marble grit floors, so cool in summer, are fiercely cold in winter, as Marcello predicted. Very much concerned for the well-being of his family, particularly that of his elder daughter, Marcello is on the phone first thing in the morning, assuring us that we will have central heating in a matter of days. The plumbing has been in place since the summer. The problem is with the convector heaters we ordered from a factory in the Veneto. A month after the due delivery date they have not arrived. But they are on their way, according to the head office. They have already been loaded onto a lorry which is at this moment heading down the Italian peninsula to Reggio Calabria. From there they will be put on the ferry to Messina and continue the journey to Palermo. Within twenty-four hours of arrival the precious load will be delivered to our door, that is a week before Christmas Day at the latest. Having given up hope of contacting the Husky plumber, we have found an alternative. A certain Signor Mirto is booked for the twenty-first. So, God willing, we will have heat for Christmas. Alleluia!

One task in which we have not been successful is the acquisition of a Christmas tree. The arrival of the German tradition in Sicily may have been delayed by a century or so but now the trees are an absolute must for one and all. The local plant vendors have sold out weeks before Christmas. The trees used here are not firs of course, but locally grown pines. Where have all the pine trees gone? To the shopkeepers of Mazara every one. Not a single shop in any of the main shopping streets is without a potted Christmas tree at the door; some sport half-hearted attempts at decoration, such as lopsided red bows and streamers, but most are not decorated at all. They just stand there in their pots looking bare and miserable. The Mazarese attitude toward Christmas trees is not very different from their attitude toward dogs—acquire one, then tie it up outside the front door and forget about it. This keeping up with the Joneses—or rather, Giacalones—fixation on the part of the Mazarese business community has seriously depleted the supply to the domestic market. But where did the rest of the family get their trees? Giusi and Rosetta, both paragons of domestic virtue, have splendid trees, so tall that they practically touch the ceiling. They ordered them weeks ago apparently. Nanette's tree is rather odd looking, however. Surely she didn't buy it? It turns out that like me, she left it too late, so she asked a local gardener to find her one. He gave her a pine branch from a tree in his garden but he didn't pay too much attention to the shape. It has two prongs at the top sticking up like a pair of horns. Even a non-Sicilian can spot the cuckold association. This *albero cornuto* will *not* make a *bella figura*. I decide to leave him out of my tree problem and wait for Marcello to deal with it.

When Marcello eventually arrives, there is a more pressing issue on his mind than Christmas trees—central heating, or rather, lack of it. Melissa and I are quite used to the cold now, dividing our time between the kitchen, Nanette's car—very warm with the heating turned up full blast—and other people's houses, which don't have central heating either, but are at least equipped with alternatives like gas stoves and fully operative fireplaces. So far there is no sign of the convector heaters, which were supposed to arrive days ago. Signor Mirto is expected tomorrow but it looks as though there will be nothing for him to install. Marcello is incensed, not so much by the prospect of a cold Christmas, but by the manufacturers' breach of faith. When he calls them he is told that the heaters have already been delivered to Palermo. He asks for the number of the depot and calls them too. Yes, the truck-load he mentions has arrived, but there is nothing for Signor Manzo. Back to the factory. He asks for the order number, then calls the depot again. Yes, that order has arrived but they were for someone else. They've just been loaded onto the depot lorry to deliver to the customer, a certain Signor Villa. More calls to the factory in the Veneto, then to Palermo. Finally he gets through to the mobile phone of the lorry driver with the controversial load. The driver reads out the address on the delivery note. He's looking for: Signor M. Villa, S. Maria, Mazara del Vallo.

Some fool has taken liberties with the punctuation. It should read "Signor Manzo, Villa Santa Maria." Having solved the conundrum, Marcello, who has been literally hopping mad all day, is now ecstatic. He rushes out and locates the poor driver, who has been cruising the local byways for some time, and escorts him triumphantly back to Santa Maria.

The long-awaited convector heaters are now stacked neatly in the warehouse ready for Signor Mirto. That evening the phone rings. Signor Mirto has bad news. He's calling from the island of Pantelleria, where he's been working on a hospital construction site for the last few weeks. The weather is so bad that no planes can take off. For the time being he's stuck on the island so he won't be turning up at Santa Maria tomorrow. He'll let us know when he's about to leave. From this point on we monitor the weather forecasts, paying particular attention to the islands in the Sicilian Channel. That night the wind drops, and we wake the next day to brilliant sunshine. Before he even raises his head from the pillow, Marcello reaches for the phone. At the other end of the line Signor Mirto is as pleased as we are about the sudden calm, but not for the same reason. Finally he gets to go home to his family after weeks of penury on this prison of an island. It may be an upmarket holiday destination, beloved of pop stars and the Milanese glitterati, but it must very well feel like a prison if you can't get off it. Marcello sympathises, then pops the question.

"So will you be coming here this afternoon?"

"I don't think so. I won't get to Mazara until this evening."

"Tomorrow?"

"What? Take off to Santa Maria on Christmas Eve after so many weeks away from home? My wife would kill me." There's no arguing with that. How had it even occurred to us that he would neglect his family just to install our heating, and on Christmas Eve of all times? The idea should not even have entered the waiting room of the mind. Marcello wishes him *Buone Feste* and heads off to Mazara in search of electric heaters.

The sun is now beaming down all warm and cheery from a cerulean sky, which is just as well under the circumstances. With the heating issue now on ice, Marcello devotes his energies to the tree problem. Unless we pinch one from outside a shop-front in the dead of night, we are not going to find a Christmas tree in Mazara. Fortunately, Maria Antonietta comes to the rescue. A large area of their land has been invaded by giant canes, which are due to be removed by a mechanical earthmover to prevent them from spreading to the rest of the property. Various *piante spontanee* have also sprung up amongst the canes on the land to be bulldozed, including a couple of beautiful pines. Emiliano, the family handyman, has removed one, roots and all and transferred it to the sitting room Next Door, where it now stands triumphant, swathed in red and green. Why doesn't Marcello take the other? Excellent idea, except that uprooting a pine tree is no mean feat for someone of his age and physique. I point this out to him but he is undeterred, and Emiliano, the only nephew who ever shows ready and willing to help out on such occasions, happens to be out. Undaunted, Marcello sets off for the cane fields armed with a spade and his gardening knife, followed by daughter, wife, two sisters-in-law, niece, Stefi, and two elderly dogs. This convoy may be a poor substitute for Emiliano in terms of muscle power, but we are ready to provide any moral support that may be required. We troop round the side of the villa, past Pilli and Maria Antonietta's house and across their land. Access to the field means wading through a sea of mud and canes twice the height of man. By the time he gets to grips with our future Christmas tree, Marcello is heavily scratched and splashed with mud. The rest of us stand and watch at a prudent distance. The object of his attentions is a magnificent specimen, at least ten feet tall. It seems a terrible shame to remove it, but guilt is allayed by the knowledge that its days are numbered anyway. In spite of the cheerleaders' encouragement, the attempt to dig up the tree is unsuccessful and leaves Marcello on his back in the mud, one knee gaping through a brand-new pair of corduroy trousers. Having failed to get the tree up by the roots, he does what we all knew he would have to do all along—cut it down. While he saws away at the trunk with his knife, expletives rain forth, mostly combinations of pig with other animals and unlikely partners such as Judas, the Madonna and misery itself: *Porca vacca, Porco cane, Porco Juda, Porca miseria,* and so forth. Still the trunk is not very thick, and it's not long till it snaps and falls to the ground, revealing an unprofessional looking jagged edge. Marcello, in he-man mode, makes as if to carry it back single-handed, but here the female militia can be of use, so we form a line and carry the thing between us, not exactly hi-hoing, but looking embarrassingly like the lineup of the Seven Dwarfs on their way back from the woods, with Jack and Lucy bringing up the rear.

Back at the house Marcello trims the tree (more sawing and swearing), fills a pot with sodden earth from the water-logged orange grove, and wedges it in with stones. Despite my protests, he insists on carrying out all these operations in the intimacy of the dining room. A week of almost constant rain had already resulted in unsightly muddy footprints on erstwhile gleaming checkerboard floors, which are now splattered with lumps of earth, pine needles, and unidentifiable vegetable matter. Having once prided myself on the Vermeer look of our *salone,* I now admit that the mess resembles nothing so much as the work of a Turner prizewinner. Lucy, who is suffering from a form of doggy dementia, does a wee for good measure, perhaps under the impression she is still in the cane field.

Still, never mind the debris underfoot and Marcello's ruined trousers, we have a splendid tree. This year I have bought color-coordinated glass baubles, so fragile they look as if they would break to the touch. My plan is to replace the lurid multicolored globes, plastic angels, satin Chinese fairies, and relics of primary school art classes accumulated over the last twenty years, with tasteful decorations in white and gold. Not all my purchases are still intact, some having come to a sad end somewhere between Sloane Square and Sicily, but there are enough survivors to do justice to the tree. To my surprise, instead of complimenting me on my good taste, Melissa is censorious and disapproving. Examining a price tag from an upmarket shop, she confronts me: "How much did you spend on all this?"

Now experiencing embarrassing Mother-Daughter role reversal, "Well, of course it didn't come cheap."

"What a waste of money. Is it supposed to be minimalist or something?"

"Well actually, I was just trying for a more harmonious theme, gold and white to match the room. *I* think it looks rather good."

"Well *I* think it looks like the kind of tree they put in hotel lobbies. Anyway Mum, we like the old decorations; they remind us of all the family Christmases in the past."

"Us? Clem's not here."

"Well if she were, she'd agree with me." Of course she would. Never for a minute would I dream of putting a finger between the Flowers. They may bicker incessantly when they're together, but when it comes to parental confrontation, they put up a united front. Finding myself cast in role of Extravagant, Childish Mother admonished by Sensible, Mature Daughter, I protest.

"Well I was only trying to do something nice."

"All right, Mum, I can see where you're coming from. But let's put the old decorations on too. We're not doing a tree for *House and Garden* you know."

THE PROLONGED BOUT OF RAIN HAS DELAYED ANY FORAYS into the grounds in search of greenery. What with all the pressure in the kitchen and the rejection of my minimalist look tree, I have left the house decoration to Melissa who, as everyone knows, is a girl of many talents. She stays up late to decorate, and the following morning we wake to find the living rooms bedecked with branches of oranges, lemons and elegant arrangements of pine cones and branches sprayed in silver and gold. She has strung out our Christmas cards over the non-functional fireplace, and by foraging through packing cases, she has found some china Nativity figures picked up in a Peruvian flea market to go on the dining room shelf. Tinsel hangs in loops from the ceiling lamps, and there, in proud defiance of the god of color coordination, stands a frenetically modified tree, the elegant glass whatnots still hanging on in there, but now completely upstaged by their gaudy predecessors. Marcello, who has been extraneous to the Christmas decoration fashion controversy, pronounces his verdict.

"E bellissimo!"

Thanks to Melissa, our household is awash with Christmas cheer. But not so the rest of Mazara. True, there are streetlights and Christmas trees all over town, but although we are thankful to be spared schmaltzy piped Santa Claus music in the shops, it still doesn't feel very Christmassy. Marcello agrees that the English version of Christmas is more festive. My theory is that to celebrate special occasions in the English way is superfluous for people who find cause for celebration whenever they get together. What a Sicilian Christmas is all about after all is mangiare.

Taking this into account, we are going to do our best to create an English Christmas lunch on Sicilian soil, a project for which Melissa and I have Marcello's full support. An English Christmas means we must have a turkey, an item that may prove hard to get, given the Mazaresi preference of fish to fowl for the feste. Over twenty years of marriage to an Englishwoman have rendered Marcello an ardent advocate of the English Christmas Lunch. In Sicilian matriarchal style, I send him to town in search of a turkey, assuming he will come back with a frozen bird shipped from the north. But I have underestimated both husband and butcher. Signor Vassallo has tracked down two live birds enjoying free-range living in the vicinity of Mazara. He's only too delighted to book one of these, a twenty-five pound male, for the Manzo table for Christmas. What about the sausages? It is not going to be easy to replicate the English meal to the letter, as non-Italian ingredients are hard to find outside the big cities, let alone provincial Sicily. As for the requisite chipolatas, the only sausages available here are the yard-long, fennel-seeded coils that Signor Vassallo makes fresh to order. Probably some sort of cross-cultural compromise will be needed. I pay him a visit to see what we can work out. First I congratulate him on nabbing one of the only two local free-range turkeys, even though it is a little on the large side. Then I ask him if he could adapt the sausage to our recipe, leave out the fennel seeds and make them smaller. Yes he could leave out the seeds and section off the sausage every two inches. No, no, Marcello interrupts. We must have the fennel seeds. All right, I concede, he may leave in the seeds, but divide it into mini-sausages please. Now, can he make the sausage thinner? Unfortunately no; he uses pig's gut to case it, and you can't mess around with the size of a pig's intestine.

"Well how do they make tiny sausages in the land where I come from?"

"I don't know, Signora. Perhaps, with all due respect, the sausage skin is not 100 percent *naturale* in your *paese*." No one would dream of eating synthetic sausage casings over here. I have so much to learn.

When the turkey gets home, he causes quite a sensation, Sicilians not being familiar with fowl so big and fat. When Lul sets eyes on the monster about to be stuffed, she lets out a shriek: *"Madonna mia, sembra un cristiano!"*

At twenty-five pounds he is indeed the size of a six-month-old baby, a thought that strikes me later as we negotiate his entry into the oven. The sausage with its little raffia bows also provokes derision. As for streaky bacon, Italian *pancetta*, which comes sliced thinly into discs instead of strips and consisting mostly of fat, will have to do. The sausage meat for the stuffing comes with compulsory fennel seeds and the bread sauce is flavored with sesame. Everything else, however, is as it should be, including the bottled cranberry sauce imported from England. Our spread would be complete, except for one article I have been unable to find: the Brussels sprout. *Cavoli di Bruxelles* are simply not in demand here, and if and when they do appear, they come from the north all packaged in plastic.

"But why have sprouts when you can have *carciofi*?" Marcello points out. Mazara is simply overflowing with artichokes. On every street corner there is a farmer's van laden with giant leafy globes, long stalks dangling over the sides like pheasants' tails.

"But these are for baking or frying," I protest. "The whole point of sprouts is to have something plain and boiled to counterbalance the richness of the rest of the meal."

"Forget it," says Marcello. "Nobody would eat them. Much better to have *carciofi alla siciliana*." I see his point. Artichokes stuffed with breadcrumbs, garlic, and pine nuts have more intrinsic appeal than boiled sprouts. It is now clear that my husband, who has been eating sprouts on Christmas Day for over twenty years with never so much as a murmur of protest, actually detests the things.

By now Pilli's family are in residence next door, and Diana, Luca, and Onda are expected for Santo Stefano, Boxing Day. Christmas Day dawns, and we rise still uncomfortably replete from the feast Nanette prepared for the *Vigilia di Natale,* the night before. With Nanette's parents also invited, there will be eighteen people around the table for lunch. The turkey has been packed tight with chestnut and sausage stuffing, coated with pancetta and butter and wrapped in aluminium foil ready for roasting. Marcello and I are up at five to put it in the oven.

Once the house is filled with the aroma of roast turkey and the kitchen is littered with the fallout from the artichoke preparation, various members of the family from Upstairs and Next Door drift in, ostensibly in order to exchange Christmas greetings. Considering that our credentials as cooks are not yet firmly established and the menu is disconcertingly unconventional, I can't help thinking they are also coming to check out the food. There are many interruptions to our preparations, as instead of sending Christmas cards, Italians tend to use the telephone to convey Christmas cheer. The phone rings constantly, first the Manzos from Trapani calling with their *auguri*, then various Vaccara cousins announcing their intentions to Get Close after lunch. Bernie has stayed up all night in order to call Melissa from New Zealand, while my family makes a joint call from my sister's home in Sussex. We all miss Clem, away from home for Christmas for the first time in her life. She'll be either fast asleep now or still celebrating Christmas Eve in the small hours of the Rio night. Maria, who is normally confined to the upstairs flat these days, is carried down the stairs by her sons for the occasion. She sits in an armchair at the end of the table, gazing in bafflement at the proceedings, while Silvio and Nanette take turns to feed her. This is the first time that all her children and grandchildren (with the exception of Clem) have been united under one roof for Christmas, but it has come too late to bring her any happiness. She seems to enjoy the food though, opening her mouth for more when the forkfuls don't come fast enough. For every one else too the lunch is a great success. The turkey lives up to Signor Vassallo's promises, and the stuffing is pronounced *eccezionale*. The most apprecia-tive of all is Anna, Nanette's mother, who is reduced to tears at the sight of turkey. She hasn't had such a treat since her last Christmas in San Diego fifteen years ago. Meanwhile the rest of the Manzos take to the novelty of the English tradition of pulling crackers and wearing funny hats. The mottoes do not go down quite so well as, even with Melissa translating, the puns come over as impossibly bad and pointless.

By the time my mother's Christmas pudding appears flaming at the table, it is almost four o'clock, and Getting Close is under way. Aldo and Lelle are the first to arrive, followed by a stream of Vaccara cousins. We soon run out of comfortable chairs, but most people are happy to sit at the dining room table, where the various dolci are set out. Perhaps sus-pecting that the Christmas pudding might not be to their taste, each family has brought supplementary supplies just in case. Maria Antoni-etta has contributed a magnificent *cassata*, while Nanette's father has brought a tray of ricotta-filled sweet pastries, confusingly named ravioli. Nanette has donated all the dolci left from the supper Upstairs the night before, and her sister has come with cannoli. I have a dauntingly huge, now beautifully decorated Christmas cake to offer, as well as meringues and trifle. But in spite of the competition, my mother's Christmas pudding steals the show. However full they may be, no Sicil-ian can resist trying a new kind of dolce, and they take tiny helpings at first, lathering on the brandy butter and whipped cream before coming back for more. I offer everyone filter coffee, apologizing that it is *tipo*

Americano and beg them to have some dolce (please, have some, please!), while Marcello hovers with the bottle of Marsala.

Conversation flows, much of it on the theme of Santa Maria and Marcello's and my role in its renaissance. Together for Christmas for the first time in nearly thirty years, the Vaccara cugini, now all in their fifties and sixties, reminisce over their childhood when they all stayed at Santa Maria and the nearby Vaccara baglio of Antalbo for the vendemmia. Some of our visitors have not set foot in the ground floor or the orange grove since they were shut off in the seventies. The sight of the newly restored house and garden evokes Proustian memories of an idyllic childhood lived in the company of a host of cousins. Wasn't it just the happiest time of their lives? Weren't they privileged to have spent their summers in such a beautiful place? Wasn't it wonderful to be part of the Family?

By late afternoon there is an alarming mountain of dolci left uneaten. As every Sicilian knows, the pastry of ravioli and cannoli quickly goes soggy. It is imperative that they are eaten the same day. What to do with these delicious but perishable dolci? Lul has a bright idea. Let's take them to the orphanage in town. She and Melissa set off on their mission of charity, only to return an hour later with the trays still laden and the mission sadly unaccomplished. Apparently no one was answering the bell at the orphange. They must have all been at church.

"Didn't you manage to get rid of any of them?"

"Yes, we did lose a few," says Melissa, "but only to a couple of fat little boys who made a grab for them as we passed them on the street."

The feasting goes on for days, from Boxing Day—Santo Stefano, Stefi's name day—through to the Feast of San Silvestro, the pious commemoration of a fourth-century pope being the Italian expression for New Year's Eve. Pilli's financial difficulties and the threat to the future of Santa Maria are never mentioned—and it is not a particularly Christmassy subject, I admit—but the days leading up to the Saint Sylvester celebration are fraught with family strife anyway. After lunch one day, when most of the Vaccara clan is gathered, Luigi Emilio voices his opinion that Silvio and Nanette are brave to contemplate inviting so many people Upstairs, especially given the precarious condition of the villa. If everyone takes to dancing, the upper floor will give way. Silvio does not agree, but the seed of doubt has been sown.

Meanwhile several people we hardly know have mentioned that they will see us "for San Silvestro," from which we have deduced that the guest list has mysteriously mushroomed. Who invited all these people? Everyone vigorously denies responsibility, but a great cloud of suspicion hangs over the younger members of Pilli's household. Apart from visions of the villa collapsing, we are now worried about security. With the gates and walls in disrepair, anyone can drive or climb into Santa Maria. How are we going to prevent the entry of gate-crashers, thieves, and drunken vandals? What was going to be simply a family get-together is getting out of control. The night before the event, Nanette and Silvio call a Family Meeting. They have decided that the party is off. There are loud protests at this, especially from Maria Antonietta.

"We can't make such a brutta figura! It's now too late to make other arrangements. We've been looking forward to this for months. The food has already been bought. You can't do this."

But Nanette puts her foot down. She is simply not having an invasion. Maria Antonietta has her head in her hands. Her greatest worry is the reaction of her consuoceri, Goffredino's future in-laws. If we cancel the party now, they will certainly be offended. The relations between the two families are seriously at risk. Goffredino's fidanzata might even call off the wedding! And then, with an accusing look at Silvio and Nanette, we'll all know who's to blame. It looks as though we could have a showdown between Pilli's family and Silvio's, with Marcello and me caught in the middle, but it doesn't happen. Tempers simmer but do not boil over, as everyone realizes that living as close as we do, we simply cannot afford to quarrel. In the end Goffredino and Emiliano undertake to sort out the security and parking arrangements, and only people on Nanette's list will be allowed into the grounds before midnight. After that, arrivals will be directed round to Next Door for the postmidnight revel. Dancing will not take place in the villa itself, and the dinner list is frozen at sixty. Nanette's mind has been put to rest, and once again the party is on.

So Saint Sylvester is celebrated with everyone on their best behavior, except for a certain seaman who turns up uninvited on the pretext that he is chaperoning one of Nanette's American friends, whose husband is away at sea. The guests have brought so much food that we can't even find enough flat surfaces to set it out on. The last guests to arrive are the *consuoceri* family, blissfully ignorant of the narrowly missed cancellation. Everything is delicious, but the star of the evening is Pilli's platter of cold baked sea bass. Apparently he brought the fish all the way from Cairo in his hand luggage, frozen of course, but tasting none the worse for the trip. In spite of the dancing ban, Pilli and Maria Antonietta take to the floor after dinner, followed by Emiliano, tall and elegant in his grandfather's morning dress, whirling the diminutive Michela across the floor. The intrusive seaman, who has so far dedicated his unwelcome attentions to Onda and Melissa, suddenly makes an ardent lunge at the unsuspecting Nanette and, under Silvio's phlegmatic gaze, sweeps her up in his arms before she can shout for help.

Fortunately, for the structure of the villa, the dancing is ballroom and brief. Before midnight we must all be down in the courtyard for a firework display orchestrated by Marcello and Massimo, who have programmed the takeoff for the stroke of twelve. We troop down the staircase, clutching bottles and champagne glasses, and out into the courtyard. The winter sky is already illuminated by a galaxy of stars. Most of us have left our coats upstairs, and we feel the air crisp against bare skin. The bell rings, and a group of young people arrive, having waited politely outside until the appropriate moment. Corks start popping. The upper terrace is alight with candles, and Marcello suddenly appears at the bell tower, ready to start the countdown. With Massimo at his side he flourishes a baton, like Ricardo Muti at La Scala: *Dieci, nove, otto, sette, sei, cinque, quattro, trè, due, uno!* A chorus of shouting breaks out, followed by a rush to kiss everyone in sight. *Auguri, auguri, auguri!* For some reason the first person to kiss me is the Signora Giuseppina.

"*Amore mio,*" she gushes, engulfing me in a fervid embrace. Meanwhile, Melissa has been claimed by her cousins. I prise her away from them, and we stand arm in arm looking up at the terrace, trying to make out where Marcello has got to. Fireworks are now shooting up, and as one shower of light descends, another explodes upward from a point farther along the roof. The sequence progresses clockwise around the top of the villa, with Marcello's and Massimo's heads just visible, gliding above the castellation as they charge on to the next launch. Down in the courtyard we are half blinded by the lights, but the conductor directs our attention to the next display with wild shouts and much waving of his stick. After all this excitement we are too tired to see the party through to the bitter end. As it turns out, only the under thirties and the uncouth seaman move on Next Door to carry on dancing till dawn. As for the Marcello and Silvio Manzos, we spend the early hours of the New Year clearing up after the last celebration of the old.

MOST PEOPLE GET SOME REST AND RELAXATION AFTER THE exertions of a New Year celebration, but I have the misfortune to have my birthday on January 2, which means that while everyone else is virtuously proclaiming their New Year Resolutions, we—Marcello and I and contingent Family—are duty-bound to prolong the eating and drinking for a further twenty-four hours. This time the indefatigable Maria Antonietta has organized a Family lunch in my honor. She has made a special trip to the marina to get seafood for *spaghetti con gli scampi.* The pasta is followed by delicious roast veal and wild asparagus, picked by my nephews from Santa Maria, then my birthday cake, a magnificent cassata.

After lunch, Marcello proposes a walk with the dogs. In competition with the mooted walk there is a film with Totò, the Neapolitan comedian, on television. Sicilians, especially the females, are not usually keen on postprandial walks. In fact, my sisters-in-law and Lul opt for Totò instead, but Silvio is morally bound to walk his dog, and Pilli, full of New Year resolutions to give up smoking, and to diet and exercise, manages to coerce his sons, even the reluctant Goffredino, to join in the walk. The sun shines clear in an impossibly blue sky, and although the temperature is over fifty degrees Fahrenheit the men wrap up with thick coats, woolly hats and scarves, as though we were setting out for the Arctic. I notice that Pilli and Emiliano stuff plastic bags into their pockets. When I ask why, they reply "You never know. We might pick up a bit of *verdura*."

Of course. The recent rains have turned the dust dry earth of summer into luxuriant forests of weeds, which, to Sicilians, are not actually weeds but potential ingredients for salad. Any excursion beyond the confines of the parco into the fields is a foraging expedition. The only weed I recognize is borage for its clumps of blue flowers.

"Why don't you pick some of that?" I ask Pilli.

"It's already in flower, which means it's past its prime. Silvio should have picked it weeks ago." Poor Silvio. He does seem to get a lot of fraternal flak over negligent husbandry of the grounds. Marcello is on the lookout for wild asparagus but Emiliano has beaten him to it, and the clumps of tender spears that lined the walls of the warehouse have already been plucked for one of Pilli's gourmet omelettes. It is Emiliano who spots the *giri*. The long stalks come easily out of the soft earth and he fills two bags in a matter of minutes.

"*Giri?* What on earth are they? Don't tell me you can eat these!"

"Of course you can."

"What are they in English?" Nobody knows. "Well, in Italian then?"

"Something like *bietole*—beet."

"Beet. Isn't that the same as Swiss chard?" The pale straggly leaves and thick roots don't look anything like the flat-leaved dark green chard you get from supermarkets. In fact giri looks like any other *erbaccia*. Pilli, who would never dream of buying verdura from a supermarket, explains that it is a *pianta rustica* and only a very distant relation of the cultivated version.

"Tonight we'll have a splendid *minestre di giri*. Or we could make an *insalata di campagna*—a field salad with broccoli, a bit of wild fennel, some capers, add olive oil and orange juice . . ."

"Or *riso con verdura*," Silvio suggests, "with asparagus and nettle leaves. Now if only we could find some tender borage leaves . . ."

Isn't it funny, I remark to Marcello, how everyone was full of New Year diet resolutions a couple of hours ago. We have only just begun to digest an enormous lunch, and here we are already discussing the next meal. But Marcello is not listening. He has strayed into the bushes beside the *concentratore*—where they once concentrated wine into must—and is bending over stones encrusted in little black shells.

"*Babaluci*—black snails! The first of the season. Emiliano, come here!" The two of them proceed to gather up the tiny crustaceans and pop them into a plastic bag. Marcello is ecstatic as he looks forward to the eating of them—in black snail soup maybe.

Having now satisfied primeval hunter-gatherer instincts, the Manzo men are loath to venture any farther across the fields, so we stick to the path that encircles the land. The sky begins to change color as we approach the Maskaro tower. A gentle jingling of bells tells us a flock of sheep is heading our way. We grab the dogs to put them on the lead before they can cause the havoc they undoubtedly have in mind. As a general rule, it is better to keep one's distance from a Sicilian shepherd, as you never know what kind of connections he may have. By law he is allowed to lead the flock into any unfenced land. In the past shepherds have presumed on this right by taking sheep right into Santa Maria to graze on the tennis court and feast on the prickly pears, which, despite the thorns, seem to be a favorite with sheep and goats.

As a result of years of confrontational incidents, relations between Santa Maria and the local shepherds are strained. As we emerge from the pine grove into our neighbor's vineyard, we come across the flock grazing away to their hearts' content. The sheep are not supposed to be here in the first place, and they should certainly not be munching away at the vines. Yet it is a truly Biblical scene, and we could be characters from a Bible story, about to berate the shepherd for leading his flock astray. If the sheep get tired of vine shoots and the shepherd decides to treat them to a feast in the parco of Santa Maria, the three Manzo brothers will surely explode into paroxysms of Old Testament wrath. By now the shepherd has seen us and is sizing up the situation. Will he defy us and let his flock leave a trail of droppings along the pine walk, then treat them to a prickly pear buffet at the villa, or will he go elsewhere? Before the brothers have time to vent the Manzo Temper, he gives way, evidently deciding he is greatly outnumbered, that is, not counting the sheep. He stands there, leaning on his stick, while the winter sun, like a giant blood orange, diffuses the landscape with a psychedelic aura of gaudy orange and red. Shards of light fall on our little gathering, casting a warm flush on olive complexions and turning the scruffy mass of woolly sheep backs into a great pink blanket. The *pastore* shouts an unintelligible salutation and heads off into the rosy-fisted sunset with his flock tinkling away behind him. We too turn tail and make our way back to the comfort of Santa Maria. Night is falling, and in the darkness we are welcomed home by the familiar call of our resident hoopoe bird.

Milton Keynes UK
Ingram Content Group UK Ltd.
UKHW041059230823
427320UK00004B/301

9 780061 373961